Advanced Engineering Surveying

problems and solutions

F A Shepherd FRICS, C Eng, FI Min E, F Inst CES

Principal Lecturer in Engineering Surveying, Trent Polytechnic

Edward Arnold

© F. A. Shepherd 1981

First published 1981
by Edward Arnold (Publishers) Ltd
41 Bedford Square, London WC1B 3DQ

All rights reserved. No part of this publication may be reproduced, stored in a retrieval system, or transmitted in any form or by any means, electronic, mechanical, photocopying, recording or otherwise, without the prior permission of Edward Arnold (Publishers) Ltd.

British Library Cataloguing in Publication Data

Shepherd, Frank Arnold
   Advanced engineering surveying.
   1. Surveying
   I. Title
   526.9'02'462   TA545

ISBN 0-7131-3416-X

Printed in Great Britain by
Thomson Litho Ltd, East Kilbride, Scotland

Preface

This book is a natural sequel to *Engineering Surveying*, published in 1977, which was based upon the first part of *Surveying Problems and Solutions*, published in 1968. Here the basic principles are applied to more advanced problems which may be encountered in Civil and Mining Engineering. As before, the main objective is to give guidance on the methods of solving typical problems by providing a discussion of the theoretical analysis of each topic, followed by worked examples and finally selective exercises for private study. The book is suitable for students in universities, polytechnics and colleges of further education who are progressing towards more applied engineering surveying syllabuses within their Degree, Higher Diploma, Higher Certificate, or professional examination courses.

The natural sequence of orientation and control measurement is followed by the theory and application of spherical trigonometry to the engineering surveyor's work. Much of the previous work on circular, transition and vertical curve alignment is updated. The material on the theory of errors and the adjustment of observation applies to so much of the surveyor's work that it is presented as a preliminary to subsequent chapters.

An attempt is made to standardise some of the notation used with the previous work and this is summarised on p. ix. Some difficulty was found in presenting a balanced view of the surveyor's and statistician's notation applied to error theory; the symbol σ has been extensively used for standard deviation and standard error when $s_{\bar{x}}$ as used by the statistician would appear to be more appropriate.

I am greatly indebted to HMSO for permission to reproduce part of their publication *Projection Tables for the Transverse Mercator Projection of Great Britain* and Table 7.1 of *Roads in Rural Areas*; to The Carriers Publishing Company Ltd for permission to reproduce Table 9 from their *Highway Transition Curve Tables*; to Wild Heerbrugg (UK) Ltd for permission to copy Figs 2.27 and 2.29 from their GAK/1 Gyro Handbook; to Dr Thomas of the Royal School of Mines for Fig. 2.26; and to AGA Geotronics for their EDM publication details.

My thanks are also due to my many colleagues at Trent Polytechnic who have given help and advice, to the many students who have wittingly or otherwise checked many of the solutions, and in particular to my son Stephen who worked long and hard at checking the original work. The ultimate responsibility for the accuracy is of course my own, and I shall always be grateful when mistakes are brought to my attention.

1981   FAS

Contents

Preface iii

Notation ix

1 Errors and their adjustment 1

1.1 Introduction 1
1.2 Terms used 1
1.3 Measures of precision 4
1.4 Uses of the terms standard error and standard deviation 6
1.5 Rejection of doubtful observations 7
1.6 Weighted observations 8
1.7 Confidence intervals (CI) 9
1.8 Propagation of systematic errors 11
1.9 Propagation of standard or random errors 12
1.10 Criterion of negligibility 13
1.11 Principles of adjustment by the method of least squares 13
1.12 Adjustment of observations 19
1.13 Adjustment by observational equations (indirect method) 19
1.14 Application to levelling circuits 26
1.15 Station adjustment 28
1.16 Adjustment by conditional equations (direct method) 32
1.17 Comparison of the indirect and direct methods 41
1.18 Adjustment by variation of coordinates (or differential displacements) 43
Bibliography 45

2 Orientation 47

2.1 Methods of orientation 47
2.2 Intersection and resection 47
2.3 Intersection 48
2.4 Multi-ray intersection 51
2.5 Weighting of physically dissimilar quantities 62
2.6 Resection 65
2.7 The direction method of triangulation 76
2.8 Satellite stations 82
2.9 The crossed quadrilateral 87
2.10 Correlation of underground and surface surveys 90
2.11 The gyroscopic theodolite 97
2.12 Alternative observational methods 100
2.13 The instrument constant (E) 106

2.14 Grid bearing of the base line 106
      Bibliography 109

3 Control measurements 111

3.1 Optical distance measurement 111
3.2 Horizontal subtense bar tacheometry 111
3.3 Factors affecting accuracy 112
3.4 Methods used in the field 115
3.5 Electromagnetic distance measurement (EDM) 119
3.6 The reduction of short and medium range EDM slope distances to the spheroid 132
3.7 Curvature of the path of the EDM wave 135
3.8 Trigonometrical heighting 139
     Bibliography 148

4 Application of spherical trigonometry to engineering surveying 149

4.1 Definitions 149
4.2 Spherical trigonometrical formulae 150
4.3 Solution of the general spherical triangle 151
4.4 Problems involving inclined planes 156
4.5 Measurements on the earth's surface 158
4.6 Convergence of the meridians 162
4.7 The spheroid 163
4.8 The transverse Mercator projection 169
4.9 Determination of approximate geographical coordinates given the grid coordinates using basic spherical trigonometry 175
4.10 Angular distortion 177
4.11 Local scale factor 178
4.12 Computation of convergence (γ) by projection tables 180
     Bibliography 183

5 Circular curves 184

5.1 Definitions 184
5.2 Through chainage 187
5.3 Length of arc (L) 188
5.4 Geometry of the curve 188
5.5 The arc/chord approximation 188
5.6 Special problems 190
5.7 Location of the curve 194
5.8 Setting out of curves 198
5.9 Computation of coordinates around the curve 205
5.10 Highway alignment design 208
5.11 Compound curves 210
5.12 Reverse curves 217

     Bibliography 222

6   Transition curves   223

6.1   Superelevation (θ)   223
6.2   Superelevation on roadways   224
6.3   Minimum curvature for standard velocity   224
6.4   Length of the transition (L)   224
6.5   Radial acceleration   225
6.6   The ideal transition curve   225
6.7   The clothoid   226
6.8   The Bernoulli lemniscate   229
6.9   The cubic parabola   229
6.10 Transition curve tables   230
6.11 Setting out processes   234
6.12 Application of superelevation on roads   236
6.13 Setting out of the transition curve where an obstruction prevents a deflection angle from the tangent point   244
6.14 Transition curves connecting compound curves   246
      Bibliography   253

7   Vertical curves   254

7.1   Properties of the simple parabola   254
7.2   Properties of the vertical curve   255
7.3   Design criteria   257
7.4   The scale factor for use of railway curves in the plotting of vertical curves to different scales   258
7.5   Sight distances   260
7.6   Setting out data   262
      Bibliography   270

Appendix   271

1   Derivatives of simple functions   271
2   Comparison of plane and spherical trigonometrical formulae   271
3   Series   272

Index   273

Notation

Greek alphabet

alpha	α	nu	ν
beta	β	xi	ξ
gamma	γ	omicron	o
delta	δ or Δ	pi	π
epsilon	ε	rho	ρ
zeta	ζ	sigma	σ or Σ
eta	η	tau	τ
theta	θ	upsilon	u
iota	ι	phi	ϕ
kappa	κ	chi	χ
lambda	λ	psi	ψ
mu	μ	omega	ω

Whilst it is almost impossible to standardise on symbols, some attempt has been made to name the most common variables.

Lengths

Plane surveying
Horizontal H
Slope length (measured) D

Geodetic surveying
Reduced length D'
Measured D
Spheroidal S
Spheroidal chord S'

Polar coordinates
Bearing A to B ϕ_{AB}
Length A to B S_{AB}
Clockwise angle $ABC = \phi_{BA} - \phi_{BC}$

Coordinates
Total Easting of A, B E_A, E_B Latitude ψ
Total Northing of A, B N_A, N_B Longitude λ

Partial coordinates of the line AB
Partial Easting of $AB = \Delta E_{AB} = E_B - E_A = S \sin \phi$
Partial Northing of $AB = \Delta N_{AB} = N_B - N_A = S \cos \phi$

Errors
Standard deviation
  of a single observation σ_x estimated s_x
Standard error
  of the mean $\sigma_{\bar{x}}$ estimated $s_{\bar{x}}$
  of the weighted mean $s_{\tilde{x}}$
Most probable value
  equal precision $v = \bar{x}$
  weighted $v_w = \tilde{x}$

Curves
Deflection angles Δ, δ
Tangential angle α
Tangent points
  on the straight T_i
  on the curve P_i
Length of arc l_i
Total length $L = \Sigma l$
Radius of arc
  minimum R
  variable on transition r
Chord c
Intersection point (IP) I

Common values used
$\pi = 3.141\,592\,654$
$e = 2.718\,281\,828$
$1_{\text{rad}} = 57.295\,779\,513° = 206\,264.806\,25'' \simeq 206\,265''$
$1/\sin 1'' = 206\,264.806\,247 \simeq 206\,265$

1
Errors and their adjustment

1.1 Introduction

For any given survey project, the work must have a specified accuracy so that the results produced may be used with confidence. For the following reasons, absolute accuracy can never be obtained.

a) Instrumental imperfections.
b) Inability of the observer to make perfect observations.
c) Variations in climatic conditions.
d) Lack of perfection in the computed or derived data.

The art of surveying is thus related to the ability of the surveyor to carry out his task to the required specification, in a minimum of time and at an economical rate. As the accuracy of the specification increases so the time required to achieve this accuracy rapidly increases.

1.2 Terms used

1.2.1 Accuracy

Accuracy is the measure of the lack of error and is an attempt to estimate the difference between the measured value and the true value. Although the absolute accuracy is never possible, **relative accuracy** may be estimated by the comparison of the measurement of the quantity with its value derived by a more accurate process. This difference, usually expressed as an error ratio (i.e. 1 in x, the ratio of the error to the size of the quantity), represents the **relative accuracy**.

1.2.2 Precision

Precision is the degree of conformity between single measurements in a series of observations.

1.2.3 Error

Error and accuracy are inversely proportional. The **true error** (E) of a single observation cannot be determined, because the **true value** is never known. **Closing errors** can present a false impression of accuracy because there may be cancellation of positive and negative errors. This is particularly likely in traversing.

1.2.4 Classification of errors

Errors may be classified as follows.

a) **Mistakes** are avoidable errors which do not conform to any law or pattern, e.g. misreading, misbooking; systematic procedures must eliminate these errors.
b) **Constant errors** are of constant magnitude and sign; e.g. standardisation of tape, vertical index error. Their effect may be minimised by systematic computation or observation.
c) **Systematic errors** are of varying magnitude but may be *either* of constant sign, e.g. malalignment of the tape, trunnion axis error; *or* of varying sign in the form of periodic errors, e.g. graduation errors. Both types are minimised by systematic observations and reduction techniques.
d) **Periodic errors** are of varying magnitude and sign but they obey systematic laws; e.g. tape and circle graduation errors. Their effects are minimised by repeated measurement on different parts of the graduations.
e) **Random** (or **accidental**) **errors** are compensating and generally unavoidable. They represent the residual errors after all others have been eliminated. They are present in all observations and are caused by imperfections in the equipment and the observer together with varying conditions. They usually conform to the laws of probability. Their effects are minimised by taking the mean of repeated observations.

1.2.5 Analysis of results

Results should be analysed using the following guidelines.

a) Constant and systematic errors are difficult to detect and checks on the results are essential, preferably using an alternative method.
b) Systematic and periodic errors are indicated by their magnitude and sign.
c) Random errors are shown by their irregular deviations from the mean.

1.2.6 Residual

The residual (r) is used in the opposite sense to error (e), i.e.

$$e = x - X \qquad (1.0)$$

where x = the measured value
and X = the true (or most probable value v)
whilst $r = X - x = v - x$ (1.1)

1.2.7 Most probable value

The most probable value (v) is that value which, based upon the observations, is more likely to be nearer to the true value than any other. Where values are measured with equal precision the **average** or **mean value** is accepted as the most probable value and is defined as

$$v = x + r \qquad (1.2)$$

1.2.8 Arithmetic mean

If a quantity x is measured n times ($n \to \infty$) under constant conditions the true arithmetic mean of the population is given as

$$\mu = \frac{1}{n}\sum_{i=1}^{n} x_i \quad \text{(as } n \to \infty\text{)} \qquad (1.3)$$

and this is assumed to be the most probable value. In practice a finite number of measurements are made and an estimate of the mean $\hat{\mu}$ is given as

$$\hat{\mu} = \bar{x} \simeq \mu = \frac{1}{n}\sum_{i=1}^{n} x_i \qquad (1.4)$$

i.e. if $x_1, x_2, x_3, \ldots, x_n$ are observed values the arithmetic mean of x is given as

$$\bar{x} = \frac{1}{n}(x_1 + x_2 + x_3 + \ldots + x_n) \qquad (1.5)$$

$$= \frac{\sum x}{n} = \frac{[x]}{n} \qquad (1.6)$$

It should be noted that in surveying the sign $\sum x$ meaning 'take the sum of' is replaced by the Gaussian notation $[x]$.

In calculating the arithmetic mean the process may be simplified by assuming a base value A and the mean value is then given as

$$\bar{x} = A + \frac{1}{n}[x - A] \qquad (1.7)$$

Example 1.1 Find the most probable value of the following angular observations: 34°26′56″, 34°27′04″, 34°26′58″, 34°26′59″, 34°27′03″, 34°27′04″.

Method 1 Summing the values quoted and dividing by six gives

$$v = \frac{1}{6}(34°\,26'\,56'' + 34°\,27'\,04'' + 34°\,26'\,58''$$
$$+ 34°\,26'\,59'' + 34°\,27'\,03'' + 34°\,27'\,04'')$$
$$= \frac{1}{6}(206°\,42'\,04'')$$

Thus $v = 34°\,27'\,00.7''$

Method 2 Let $A = 34°\,27'\,00''$. Then

$$v = 34°\,27'\,00'' + \tfrac{1}{6}(-4+4-2-1+3+4)''$$
$$= 34°\,27'\,00'' + 00.7''$$
$$= 34°\,27'\,00.7''$$

Method 3 Let $A = 34°\,26'\,56''$ (i.e. the smallest value). Then

$$v = 34°\,26'\,56'' + \tfrac{1}{6}(0+8+2+3+7+8)''$$
$$= 34°\,26'\,56'' + 4.7''$$
$$= 34°\,27'\,00.7''$$

Method 3 is probably the best as the mean adjustment is always positive.

1.3 Measures of precision

The distribution of random errors is assumed to be *normal* and obeys the following rules.

a) Small errors are more frequent than large errors.
b) Positive and negative errors occur with equal frequency when $n \to \infty$, i.e. they are compensating.
c) Very large errors do not occur.
d) They conform to the laws of probability, i.e. the probability (y) that a residual error (r) will occur is given by

$$y = h\pi^{-1/2}e^{-h^2r^2} \qquad (1.8)$$

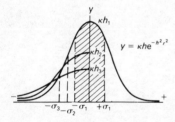

Fig. 1.1

where h is the index of precision and e is exponential e (see Fig. 1.1).

The following points should be noted.

i) As $r \to \pm\infty$, $y \to 0$.
ii) As $r \to 0$, $y \to$ a maximum of $h\pi^{-1/2}$.
iii) The sum of all the probabilities of r is given by

$$\int_{-\infty}^{\infty} y = 1 \qquad (1.9)$$

i.e. the probability of $r(-\infty \leqslant r \leqslant \infty)$ is certain, i.e. 1 is the total area under the curve.

iv) As h increases, y_{max} increases.
v) For a given set of measurements, the more precise the measurement the more chance of the residuals tending to zero.

Referring to Fig. 1.1, the value of y at any point on the curve gives the probability of the occurrence of r. The area under the curve between the two values of r, say $\pm\sigma_1$, is the sum of the probabilities of occurrence of the residuals with values $-\sigma \leqslant r \leqslant +\sigma$ and can be shown to be 68.3% of the total area under the curve. It can also be shown that the percentage area for a multiple of σ is fixed.

The probability of occurrence may be expressed as (z), the number of standard deviations from the mean (see p. 9).

1.3.1 Variance

The measure of precision is known as the variance (σ^2), which represents the theoretical mean of the squares of the errors.

$$\sigma^2 = [(x_i - \mu)^2]/n \tag{1.10}$$

1.3.2 Standard deviation

The standard deviation (σ) is defined as the positive square root of the variance, i.e.

$$\sigma = \left(\frac{[(x_i - \mu)^2]}{n}\right)^{1/2}_{n \to \infty} \quad \text{for } i = 1 \text{ to } n \tag{1.11}$$

1.3.3 Sample precision

The term 'sample' implies a limited series of observations within the population of the variables. As the theoretical mean μ is given as $\bar{x}$, an estimate of the variance has to be made, and is given as

$$s_x^2 = \left(\frac{[(x_i - \bar{x})^2]}{n - 1}\right) = \frac{[r_i^2]}{n - 1} \tag{1.12}$$

Note that as the theoretical mean is not known the estimate $\bar{x}$ is used. Also, because the number of observations is small (in surveying usually less than 30) n is replaced by $(n-1)$, the number of degrees of freedom.

In surveying, the term **standard error** is commonly used as synonymous with standard deviation but by definition in BS 2846 (1957), 'it is generally reserved for the standard deviation of the mean.'

In statistical terms standard error is given as

$$SE = \sigma/\sqrt{n} \tag{1.13}$$

and thus a similar term in surveying is used as

$$s_{\bar{x}} = \frac{s_x}{\sqrt{n}} = \left(\frac{[r_i^2]}{n(n-1)}\right)^{1/2} \tag{1.14}$$

1.4 Uses of the terms standard error and standard deviation

Standard error in surveying is frequently used to define the accuracy of observed or computed values. Thus a line is said to be, say, 126.352 ± 0.002, the quantity ± 0.002 representing the standard error of the most probable value, i.e. the mean. It is seldom that single observations are taken and the standard deviation of a single observation (s_x) is used only for rejection criteria (see p. 7).

If in certain circumstances, say using a theodolite which has been in use over a long period, the standard deviation (σ) may be known and used as such. But in most cases only a sample estimate of the standard deviation can be derived and then the standard error ($s_{\bar{x}}$) should be used.

An alternative method of computing the standard deviation, more suitable for machine computation, is to use the following expression:

$$s_x = \left(\frac{[x^2] - [x]^2/n}{n-1}\right)^{1/2} \tag{1.15}$$

Example 1.2 A baseline was measured in metres, ten times, with the following results: 126.342, 126.349, 126.351, 126.345, 126.348, 126.350, 126.348, 126.352, 126.345, 126.348. Calculate (a) the most probable value, (b) the standard deviation and (c) the standard error of the mean.

(a) The most probable value is equal to the arithmetic mean, i.e.

$v = \bar{x} = 126.3400\,\text{m}$
$\qquad + (2 + 9 + 11 + 5 + 8 + 10 + 8 + 12 + 5 + 8)/10\,\text{mm}$
$\qquad = 126.3400\,\text{m} + 0.0078\,\text{m}$
$\qquad = 126.3478\,\text{m}$

(b) Using equation (1.15), the standard deviation is given by

$$s_x = \left(\frac{[x^2] - [x]^2/n}{n-1}\right)^{1/2} = \left(\frac{692.00 - (78.00^2/10)}{9}\right)^{1/2}$$

$\qquad = \pm 3.05\,\text{mm}$

(c) The standard error is given by

$s_{\bar{x}} = s_x/\sqrt{n} = \pm 3.05/\sqrt{10}$
$\qquad = \pm 0.96\,\text{mm}$

1.5 Rejection of doubtful observations

A series of observations may contain one value which lies outside the normal distribution and this must be rejected. In practical terms it is suggested that where $n > 10$, the residual of any observation should not exceed $2.5s_x$ whilst for $n < 10$, the multiplying factor should range from 2.5 to $3.5s_x$, giving a confidence level of 99% (see p. 9).

A field value for the standard error of the mean ($s'_{\bar{x}}$) is given as Peters' approximation

$$s'_{\bar{x}} = \frac{5\,r'}{4\sqrt{(n-1)}} \tag{1.16}$$

When $n = 10$,

$$s'_{\bar{x}} = 5\,r'/12 \tag{1.17}$$

and $\quad s_x = \sqrt{n} \times s'_{\bar{x}} = \sqrt{10} \times 5r'/4 \simeq 4r'/3 \tag{1.18}$

Example 1.3 The following subtense bar observations (see also Chapter 3) were recorded as follows.

	r	r^2	New r
2° 31′ 15″	0.3	0.09	0.4
2° 31′ 14″	0.7	0.49	1.4
2° 31′ 08″	6.7	44.89	rejected
2° 31′ 15″	0.3	0.09	0.4
2° 31′ 17″	2.3	5.29	1.6
2° 31′ 15″	0.3	0.09	0.4
2° 31′ 16″	1.3	1.69	0.6
2° 31′ 15″	0.3	0.09	0.4
2° 31′ 17″	2.3	5.29	1.6
2° 31′ 15″	0.3	0.09	0.4
147″	14.8	58.10	7.2

Thus $\bar{x} = v = 14.7''\quad 1.48(r')\quad 0.72(r')$

Then $\quad s_x = \sqrt{(58.10/9)} = \pm 2.54''$

and $\quad s'_{\bar{x}} = 4 \times 1.48/3 = \pm 1.97$ (rather low)

Rejection of $2.5 s'_x = 2.5 \times 2.5 = 6.25''\quad$ (3rd entry rejected)

New mean = 15.4″

$s_{\bar{x}} = \pm 0.34''\quad (s'_{\bar{x}} = 5r'/(4\sqrt{8}) = \pm 0.32'')$
$s_x = \pm 1.02''\quad (s'_x = 0.32 \times \sqrt{8} = \pm 0.91'')$

Accepting a confidence level of $z = 2.5(s_{\bar{x}})$ (i.e. 98.8%), then

$\alpha = 2° 31' 15.4'' \pm (2.5 \times 0.34)'' = 2° 31' 15.4'' \pm 0.85''$
$\alpha/2 = 1° 15' 37.7'' \pm 0.42''$

The horizontal length is then cot $(\alpha/2)$, and assuming the bar is 2 m this is equal to

$$45.448 \text{ m} \pm 0.004 \text{ m} \quad (\text{error ratio} \simeq 1/11\,360)$$

Note that the standard error in the length is obtained by including the change of 0.42" in the value of cot $1°\,15'\,37.7"$.

1.6 Weighted observations

Where observations or derived values are not of equal precision, then a **weight** may be attributed to the values. This weight is a measure of the reliability of one value compared with others; the better the precision, the higher the weight. Weight (w) may be applied as follows.

a) w may be in proportion to the number of observations assuming equal precision, i.e.

$$w \propto n \tag{1.19}$$

It should be noted that n may be assumed to represent the frequency of the occurrence of a value within a series.

b) w may be inversersly proportional to the square of the standard error of the observations or of the derived value.

As $\quad s_{\bar{x}} = s_x/\sqrt{n}$

$$= s_x/\sqrt{w} \tag{1.20}$$

then $\quad w \propto 1/(s_{\bar{x}})^2 \tag{1.20a}$

With the observations $x_1, x_2, \ldots, x_n$, having weights $w_1, w_2 \ldots, w_n$ respectively, the most probable value, known as the **weighted mean**, is given as

$$\tilde{x} = \frac{w_1 x_1 + w_2 x_2 + \ldots + w_n x_n}{w_1 + w_2 + \ldots + w_n} = \frac{[w_i x_i]}{[w_i]} \tag{1.21}$$

The weight of the weighted mean is thus $[w_i]$.

Cooper suggests that if we let $w_i \sigma_i^2 = \sigma_0^2$ (called 'the variance of a measurement of unit weight') then the variance of the weighted mean becomes

$$s_{\bar{x}}^2 = \sigma_0^2/[w_i] \tag{1.22}$$

Thus an unbiased estimate is given as

$$s_0^2 = [w_i(x_i - \tilde{x})^2]/(n-1)$$

where $s_0^2 \simeq \sigma_0^2$. Then an unbiased estimate of the variance of an observation x_i of weight w_i is given as

$$s_{xi}^2 = \frac{s_0^2}{w_i} = \frac{[w_i(x_i - \tilde{x})^2]}{w_i(n-1)} \tag{1.23}$$

This is usually known as the **standard error of an observation of** weight w_i. Similarly, the **standard error of the weighted mean** is given as

$$s_{\bar{x}}^2 = \frac{s_0^2}{[w_i]} = \frac{[w_i(x_i - \hat{x})^2]}{[w_i](n-1)} \qquad (1.24)$$

1.7 Confidence intervals (CI)

1.7.1 CI for theoretical mean

In a series of observations in which the errors are said to be normally distributed, a confidence interval for the theoretical mean (μ) of $100(1-\alpha)\%$ is given as

$$\bar{x} \pm z_{\alpha/2} \sigma_{\bar{x}} \qquad (1.25)$$

where the standard normal variable is given by

$$z = (x - \mu)/\sigma \qquad (1.26)$$

and $(1-\alpha)$ is the degree of confidence; i.e. it is $100(1-\alpha)\%$ certain that $\bar{x}$ is not more than $\pm z_{\alpha/2} \sigma_{\bar{x}}$ away from the theoretical mean. For example, if $(1-\alpha) = 0.95$, this gives 95% CI. From a table of distribution values,

$$z_{\alpha/2} = z_{0.025} = 1.96$$

Other values frequently used are shown in the following table.

CI %	z
50.00	0.6745
68.27	1.00
95.55	2.00
98.00	2.33
98.80	2.50
99.74	3.00

1.7.2 Student-t distribution

In the early part of this century W. S. Gossett, under his pen-name 'Student', developed a theory for small samples and in particular a formula for a sampling distribution of the mean value when (σ) is not known and an estimate (s) has to be made. This is known as the 'Student-t' distribution which, as the sample size decreases, becomes flatter and wider than the normal distribution. The value of t is such that $ts/\sqrt{n}$ will not exceed the given probability. Thus

$$t = \frac{(\bar{x} - \mu)}{(\sigma/\sqrt{n})} = \frac{\text{the true error of the mean}}{\text{the standard error of the mean}} \qquad (1.27)$$

A new confidence interval of $100(1-\alpha)\%$ is now given as

$$\bar{x} \pm t_{\alpha/2}\, s/\sqrt{n} \qquad (1.28)$$

where $t_{\alpha/2}$ is the 'Student-t' value given in tables with $(n-1)$ degrees of freedom.

The value of $t_{\alpha/2}$ given in the table below has a probability of $\alpha/2$ being exceeded for different values of n, with $(n-1)$ degrees of freedom.

1.7.3 Table of 'Student-t' values for selected values of n

n	95% ($t_{0.025}$)	98% ($t_{0.01}$)	99% ($t_{0.005}$)
4	3.18	4.54	5.84
6	2.57	3.37	4.03
8	2.37	3.00	3.50
9	2.31	2.90	3.36
10	2.26	2.82	3.25
12	2.20	2.72	3.11
16	2.13	2.60	2.95
17	2.12	2.58	2.92
18	2.11	2.57	2.90
19	2.10	2.55	2.89
20	2.09	2.54	2.86

1.7.4 Calculation of the sample size

To be $100(1-\alpha)\%$ certain that $\bar{x}$ is not more than a specified amount (L) away from the theoretical mean (μ),

$$L = t_{\alpha/2} s/\sqrt{n} \qquad (1.28a)$$

e.g. if $100(1-\alpha)\% = 95\%$

then $\alpha/2 = 0.05/2 = 0.025$

and if the specified range is

$$L = t_{0.025}\, s/\sqrt{n}$$

then $\quad n = (t_{0.025}\, s/L)^2$

is the number of observations in the sample.

Example 1.4 The application of weights from frequency.
Either of the following two methods may be used.

1.8 PROPAGATION OF SYSTEMATIC ERRORS

x	r	r²		x	w	wx	r	r²	wr²
268	1	1		268	3	804	1	1	3
266	1	1		266	2	532	1	1	2
268	1	1		265	3	795	2	4	12
265	2	4		269	1	269	2	4	4
266	1	1		270	1	270	3	9	9
265	2	4			10	2670			30
269	2	4							
270	3	9							
265	2	4							
268	1	1							
2670	16	30							

$v = \bar{x} = 2670/10 = 267$

$s_x = \sqrt{(30/9)} = \pm 1.83$

$s_{\bar{x}} = s_x/\sqrt{n} = \pm 1.83/\sqrt{10}$
$= \pm 0.58$

$s'_{\bar{x}} = 5 \times 1.6/12 = \pm 0.67$

$s'_x = \sqrt{10} \times \pm 0.67 = \pm 2.12$

$v_w = \tilde{x} = [wx]/[w]$
$= 2670/10 = 267$

$s_{\tilde{x}} = \sqrt{\{30/(10 \times 9)\}}$
$= \pm 0.58$

$s_{x_i} = \sqrt{\{[wr^2]/w_i(n-1)\}}$

For w_i, where $i = 1$ or 3,
$s_{x_1} = \sqrt{\{30/(1 \times 9)\}}$
$= \pm 1.83$
$s_{x_3} = \sqrt{\{30/(3 \times 9)\}}$
$= \pm 1.06$

1.8 Propagation of systematic errors

When the errors are of known value and sign, their effect may be evaluated as follows. If

$$y = f(x_1, x_2, x_3, \ldots, x_n)$$

and each variable is subjected to systematic errors Δx_1, Δx_2, Δx_3, ..., Δx_n respectively, then

$$\Delta y = \frac{\partial y}{\partial x_1}\Delta x_1 + \frac{\partial y}{\partial x_2}\Delta x_2 + \frac{\partial y}{\partial x_3}\Delta x_3 + \ldots + \frac{\partial y}{\partial x_n}\Delta x_n$$

(1.29)

i.e. the function is partially differentiated with respect to each of the variables and summed as above.

Example 1.5 If in Fig. 1.2 $L = d \cos \theta$, where systematic errors in α and θ are Δd and $\Delta \theta$ respectively, then

$\Delta L = (\cos \theta) \Delta d + (-d \sin \theta) \Delta \theta$
$= L(\Delta d/d) + L(-\tan \theta \Delta \theta)$
$= L[(\Delta d/d) - (\tan \theta \Delta \theta)]$

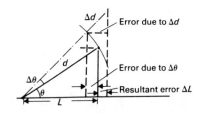

Fig. 1.2

The error ratio is then

$$\frac{\Delta L}{L} = \frac{\Delta d}{d} - \tan\theta \Delta\theta$$

i.e. the terms represent the effects of each systematic error on the resultant error ratio.

Given $d = 30.000$ m with a systematic error of $+0.010$ m
and $\theta = 5°\,00'\,00''$ with a systematic error of $-20''$
then $L = 30.000 \cos 5.000° = 29.886$ m
and $\Delta L = 29.886\,[(0.010/30.000) - (0.08749 \times -20/206\,265)]$
$= 29.886\,(0.000\,33 + 0.000\,01)$
$= 0.010\,26$

i.e. 0.010 to be compatible with the measured data. Then

error ratio $= 0.010\,26/29.886 = 0.000\,34$

i.e. the error ratio $\simeq 1/2912$.

Note that it can be seen that the effect of the angular error here is negligible.

1.9 Propagation of standard or random errors

When observations are combined to form one result, the latter has a standard error equal to the square root of the sum of the squares of the individual standard errors, i.e. if

$$x = x_1 + x_2 + x_3 + \ldots + x_n$$

then $\sigma_x = \sqrt{(\sigma_1^2 + \sigma_2^2 + \sigma_3^2 + \ldots + \sigma_n^2)}$ (1.30)

If $y = f(x_1, x_2, x_3, \ldots, x_n)$, where $x_1, x_2, x_3, \ldots, x_n$ are independent values with $\pm\delta x_1, \pm\delta x_2, \pm\delta x_3, \ldots, \pm\delta x_n$ respectively, it can be shown that by the use of Taylor's theorem, where the errors are small and the product terms are neglected, the total error in y is given as

$$\delta y^2 \simeq \left(\frac{\partial f}{\partial x_1}\right)^2 \delta x_1^2 + \left(\frac{\partial f}{\partial x_2}\right)^2 \delta x_2^2 + \left(\frac{\partial f}{\partial x_3}\right)^2 \delta x_3^2 + \ldots$$
$$+ \left(\frac{\partial f}{\partial x_n}\right)^2 \delta x_n^2 \quad (1.31)$$

The error δx etc. may be the standard deviation (σ), the standard error of the mean ($\bar{\sigma}$) or some estimated value of the error in x.

This combination of errors may be illustrated as follows. Suppose

$$a = b \sin A / \sin B$$

with errors δb, δA and δB. Partially differentiating the equation with respect to each of the variables in turn, as above, gives

$$\delta a^2 = (\sin A \, \delta b/\sin B)^2 + (b \cos A \, \delta A/\sin B)^2$$
$$+ (b \sin A \cos B \, \delta B/\sin^2 B)^2$$
$$= a^2[(\delta b/b)^2 + (\cot A \, \delta A)^2 + (\cot B \, \delta B)^2]$$

and the error ratio is given by

$$\delta a/a \simeq [(\delta b/b)^2 + (\cot A \, \delta A)^2 + (\cot B \, \delta B)^2]^{1/2}$$

1.10 Criterion of negligibility

Where standard errors are propagated in the form $z = \sqrt{(x^2 + y^2)}$, it may be that y is so much smaller than x as to render it negligible. Briggs states that when adding two standard errors of which the first is three times the latter, then the latter may be considered negligible. For example, if

$$y = x/3$$
then $\quad z = \sqrt{(x^2 + x^2/9)} = \pm 1.05 x$

As the standard errors can seldom be assessed to within 5% then the effect of y is considered negligible.

1.11 Principles of adjustment by the method of least squares

Consider the probability density equation

$$y = Khe^{-h^2 r^2}$$

To obtain the maximum value of the precision index h, the equation is differentiated and equated to zero, i.e.

$$dy/dh = K[e^{-h^2 r^2} + h(-2hr^2 e^{-h^2 r^2})] = 0$$
$$dy/dh = K e^{-h^2 r^2}(1 - 2h^2 r^2) = 0$$

Then $\quad r^2 = 1/(2h^2)$
$$h = 1/\sqrt{(2r)} \qquad (1.32)$$
$$[r^2] = 1/(2h_1^2) + 1/(2h_2^2) + 1/(2h_3^2) + \ldots + 1/(2h_n^2)$$
$$[r^2] = [1/(2h^2)] \qquad (1.33)$$

As h increases the residual errors decrease, i.e. $[r^2]$ must be a minimum when $[1/(2h^2)]$ is a maximum.

The principle of least squares may thus be stated as follows. The most probable value is obtained when the sum of the squares of the residual errors is a minimum.

The arithmetic mean conforms to this rule. If

$$r_i = v - x_i$$
and $\quad P = [r^2] = nv^2 - 2v[x] + [x^2]$

then for a minimum value of P,
$$dP/dv = 2nv - 2[x] = 0 \quad (d^2P/du^2 = 2n > 0)$$
so $\quad v = [x]/n = \bar{x}$

Example 1.6 Two observers A and B measure the same quantity d five times each with the following recorded data.

    A   149.75, 149.78, 149.83, 149.80, 149.74

    B   149.80, 149.82, 149.78, 149.70, 149.75

Calculate the most probable value of the quantity and its standard error.

	A			B	
d	r	r^2	d	r	r^2
149.75	3	9	149.80	3	9
149.78	0	0	149.82	5	25
149.83	5	25	149.78	1	1
149.80	2	4	149.70	7	49
149.74	4	16	149.75	2	4
$v_a = 149.78$		54	$v_b = 149.77$		88

Thus

$s_{\bar{x}a} = \sqrt{(54/20)} = \pm 1.64$ and $s_{\bar{x}b} = \sqrt{(88/20)} = \pm 2.10$

but $w \propto 1/s_{\bar{x}}^2 = k/s_{\bar{x}}^2$ and $s_{\bar{x}} = k/\sqrt{w}$

so $w_a \propto 1/1.64^2$ and $w_b \propto 1/2.10^2$

or $w_a = 1/1.64^2 = 0.372$ and $w_b = 1/2.10^2 = 0.227$

(when $k = 1$).

Then $\quad v_w = \tilde{x} = \dfrac{(w_a \times v_a) + (w_b \times v_b)}{w_a + w_b}$

$\qquad\qquad = 149.000 + \dfrac{(0.372 \times 0.78) + (0.227 + 0.77)}{0.372 + 0.227}$

$\qquad\qquad = 149.000 + 0.776 = 149.776$

Thus $\quad w_{\tilde{x}} = 0.372 + 0.227 = 0.599$

and $\quad s_{\tilde{x}} = 1/\sqrt{w_{\tilde{x}}} = 1/\sqrt{0.599} = \pm 1.29$

Example 1.7 Given a theodolite with a known standard error $\pm 2''$ of a single observation (σ) under given conditions, it is required to measure an angle θ to $\pm 1''$ with 95% confidence. Determine the number of observations needed.

(a) The confidence limits for a normal distribution in which

(σ) is known to be $\pm 2''$ is given as follows for $z = 1.96$ (equation 1.20).

$$\delta\theta = 1.96 \, \sigma/\sqrt{n} = \pm 1''$$
$$n = (1.96 \times 2)^2 \quad \text{say } 16$$

(b) If the $\pm 2''$ is only an estimate (s_x) then by equation 1.28a,

$$L = \delta\theta = t_{\alpha/2} \, s_x/\sqrt{n} = 1'' = t_{0.025} \times 2/\sqrt{n}$$

Then $n = (t_{0.025} \, s_x/L)^2 = (t_{0.025} \times 2/1)^2$

With reference to the tables (p. 10), when $t_{\alpha/2} = 2.10$, $n = 19$, whilst the equation gives $n = 17.6$. When $t_{\alpha/2} = 2.11$, $n = 18$, whilst the equation gives $n = 17.8$ (a better approximation).

Example 1.8 If a line is to be measured to an accuracy of ± 3 mm, when the line is approximately 30 m long and its angle of inclination is $5°\,00'\,00''$, calculate the allowable error in the angle.

Given $H = L \cos \alpha$

then by the calculus applied to errors,

$$\delta H = -L \sin \alpha \, \delta\alpha$$

Then $\delta\alpha = \delta H/(-L \sin \alpha)$

The allowable error is

$$\delta\alpha'' = \frac{206\,265 \times \pm 0.003}{30 \sin 5.00°} = \pm 237'' = \pm 0°\,03'\,57''$$

Example 1.9 To calculate the equivalent error allowable in levelling the two ends of the line in Example 1.8,

Given $d = L \sin \alpha = 30 \sin 5.00° = 2.615$ m
and $c = d^2/(2L)$

then $\delta c = \dfrac{2d\,\delta d}{2L} = \dfrac{2.615 \times \delta d}{30} = \pm 0.003$ m

The allowable error is

$$\delta d = 30 \times \pm 0.003/2.615 = \pm 0.034 \text{ m}$$

To check:

$$\delta d = L\,\delta\alpha = \frac{30 \times \pm 237}{206\,265} = \pm 0.034 \text{ m}$$

Example 1.10 In adjacent triangles ABC and CBD, with BC the common side (see Fig. 1.3), the following clockwise angles were measured: $BAC = 60°\,24'\,32'' \pm 5.5''$; $ACB = 50°\,14'\,10'' \pm 3.3''$; $DBC = 100°\,26'\,30'' \pm 2.5''$; $CDB = 40°\,20'\,25'' \pm 3.0''$. The length of the line AB was measured as 8686.80 m ± 0.05 m. Compute the length of the line CD and its resultant error.

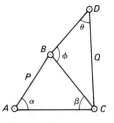

Fig. 1.3

ERRORS AND THEIR ADJUSTMENT

Let
$$AB = P = 8686.80 \pm 0.05 \text{ m}$$
$$BAC = \alpha = 60° 24' 32'' \pm 5.5''$$
$$ACB = \beta = 50° 14' 10'' \pm 3.3''$$
$$DBC = \phi = 100° 26' 30'' \pm 2.5''$$
$$CDB = \theta = 40° 20' 25'' \pm 3.0''$$

Then $BC = P \sin \alpha / \sin \beta$

$$DC = Q = BC \sin \phi / \sin \theta = \frac{P \sin \alpha \sin \phi}{\sin \beta \sin \theta}$$
$$= 14\,929.34 \text{ m}$$

Partially differentiating $f(Q)$ with respect to each of the variables and writing in terms of the errors $\partial Q / \partial P \equiv \delta Q_p / \delta P$, we have

$$\delta Q_P = \frac{\sin \alpha \sin \phi \, \delta P}{\sin \beta \sin \theta} = Q \, \delta P / P$$
$$= \frac{14\,929.34 \times \pm 0.05}{8686.80} = \pm 0.086 \text{ m}$$

$$\delta Q_\alpha = \frac{P \cos \alpha \sin \phi \, \delta \alpha}{\sin \beta \sin \theta} = Q \cot \alpha \, \delta \alpha$$
$$= \frac{14\,929.34 \times 0.567\,87 \times \pm 5.5}{206\,265} = \pm 0.226 \text{ m}$$

Similarly

$$\delta Q_\phi = Q \cot \phi \, \delta \phi$$
$$= \frac{14\,929.34 \times -0.184\,29 \times \pm 2.5}{206\,265} = \pm 0.033 \text{ m}$$

$$\delta Q_\beta = \frac{-P \sin \alpha \sin \phi \cos \beta \, \delta \beta}{\sin^2 \beta \sin \theta} = -Q \cot \beta \, \delta \beta$$
$$= \frac{-14\,929.34 \times 0.832\,10 \times \pm 3.3}{206\,265} = \pm 0.199 \text{ m}$$

Similarly

$$\delta Q_\theta = -Q \cot \theta \, \delta \theta$$
$$= \frac{-14\,929.34 \times 1.177\,48 \times \pm 3.0}{206\,265} = \pm 0.256 \text{ m}$$

The resultant error is then the RMS, given by

$$\text{RMS} = \sqrt{(0.086^2 + 0.226^2 + 0.033^2 + 0.199^2 + 0.256^2)}$$
$$= \pm 0.406 \text{ m}$$

Therefore $Q = 14\,929.34 \pm 0.41 \text{ m}$.

Exercises 1.1

1 Eight measurements of a line yielded the following values: 1820.00 m +0.50, 0.61, 0.51, 0.50, 0.40, 0.50, 0.50, and 0.47. Are any of the measurements capable of being rejected? Find the most probable value. (Ans. No rejection. 1820.50 m)

2 The following mean values were recorded for angles in a triangle ABC, the individual measurements being of equal precision.

A	62° 08′ 13″	6 observations
B	56° 42′ 30″	4 observations
C	61° 09′ 12″	2 observations

Calculate the most probable values of the angles.
(Ans. 62° 08′ 13.9″, 56° 42′ 31.4″, 61° 09′ 14.7″)

3 The sides of a rectangle were measured as 4.66 m and 3.11 m with standard errors of ± 30 mm and ± 15 mm respectively. Calculate the standard errors in (a) the calculated perimeter, and (b) the calculated area. (Ans. 0.067 m; 0.12 m^2)

4 A base line of 10 bays was measured by a tape resting on measuring heads. One observer read one end whilst the other observer read the other—the difference in readings giving the observed length of the bays. Bays 1, 2 and 5 were measured six times; bays 3, 6 and 9 were measured five times, and the remaining bays four times, the means being calculated in each case.

If the standard errors of the single readings by the two observers were known to be 1 mm and 1.2 mm, what will be the standard error in the whole line due only to reading errors?
(Ans. ± 2.3 mm)

5 The following observations were made on a test line.
363.263 (wt. 1), 363.265 (wt. 2), 363.268 (wt. 2),
363.264 (wt. 1), 363.267 (wt. 2), 363.270 (wt. 1),
363.265 (wt. 2), 363.262 (wt. 1), 363.270 (wt. 1),
363.267 (wt. 2).

Compute the most probable value and the standard error of: (a) a single observation of unit weight; (b) a single observation of weight two; (c) the weighted mean.
(Ans. 363.266 m; ± 2.9 mm; ± 2.1 mm; ± 0.8 mm)

6 The following measurements of one angle were made with two different observers using the same instrument.

A	B
43° 51′ 10″	43° 50′ 40″
43° 50′ 40″	43° 50′ 25″
43° 50′ 30″	43° 50′ 40″
43° 51′ 00″	43° 51′ 15″

Compare the reliability of the two observers and calculate the most probable value of the angle and its standard error.

(Ans. 43° 50′ 47.9″ ± 6.9″)

7 The slope length of a line AB was recorded as 126.343 m at a measured inclination of $+10° 26′ 30″$. The length was subsequently adjusted by a value of $+30$ mm and the angle by $+15″$.
(a) Compute the horizontal length of the line based upon the original observation values and the effect that the adjusted values have on the result.
(b) Assuming that the adjustment values represent standard errors, find the resultant standard error on the horizontal length. (Ans. 124.251 m; 0.028 m; ±0.030 m)

8 A target T is set on a dam face and is intersected by the horizontal angles measured at each end of a base line AB, the length of which was measured as 654.367 m with a standard error of ±2.3 mm. The observed angles and their standard errors are recorded as

BAT 73° 14′ 16.8″ ± 1.2″
TBA 59° 29′ 13.5″ ± 0.8″

Calculate the standard errors in the coordinates of T, given that A has local coordinates (**0, 0**) and the bearing of AB is due E.

(Ans. ±5 mm E; ±6 mm N)

9 An EDM instrument attached to a theodolite is to be used to set out a point P from a coordinated base line AB. It is thought that the use of one face only is necessary for the horizontal angle and tests show that the combined instruments and observational errors is ±01′ 00′. Tests on reduced horizontal lengths show that the standard error is ±0.010 m.
 Given the recorded data shown below calculate the coordinates of P and their standard errors.

	E (m)	N (m)
A	1000.000	1000.000
B	1236.432	1542.386

Theodolite station A
  Pointing to B Horizontal circle reading   10° 30′ 00″
  Pointing to P Horizontal circle reading   25° 45′ 00″
              Vertical zenith angle   86° 25′ 45″
Measured slope length AP 240.818 m
(RICS/M Ans. 1150.61 m E ± 0.06 m, 1187.31 m N ± 0.04 m)

1.12 Adjustment of observations

The most probable values of measured quantities, subject only to random errors, may be derived such that the sum of the squares of the adjustments (i.e. the residual errors) is a minimum.

It is common practice to obtain **redundant observations** in order that mistakes are detected and with the additional measurements, better estimates of the most probable values are made.

The most probable values of the unknowns may be found by

a) observational equations, i.e. the indirect method, *or*
b) conditional equations, i.e. the direct method.

1.13 Adjustment by observational equations (indirect method)

Where the number of observed values is greater than the number of unknown values, the application of the principle of least squares produces a set of **normal equations** as follows.

Given a set of unknown quantities, X, Y and Z, in the form of an equation in which a measured value $M_1, M_2, \ldots, M_m$ is subjected to a residual error $r_1, r_2, \ldots, r_m$, etc. then the observational equations may be written as follows.

$$\left. \begin{array}{l} a_1 X + b_1 Y + c_1 Z = M_1 + r_1 \quad \text{weight } w_1 \\ a_2 X + b_2 Y + c_2 Z = M_2 + r_2 \quad \text{weight } w_2 \\ a_3 X + b_3 Y + c_3 Z = M_3 + r_3 \quad \text{weight } w_3 \\ \quad \vdots \quad \quad \vdots \quad \quad \vdots \quad \quad \quad \vdots \quad \quad \vdots \\ a_m X + b_m Y + c_m Z = M_m + r_m \quad \text{weight } w_m \end{array} \right\} \quad (1.34)$$

As the values X, Y and Z may be large unwieldy values, then it is usual to substitute assumed values, e.g. X_0, Y_0 and Z_0 such that $X = X_0 + x$, $Y = Y_0 + y$ and $Z = Z_0 + z$, and the new equations may then be written in general terms as

$$a_i x + b_i y + c_i z = l_i + r_i \quad (w_i)$$

The observational equations may now be written in general terms as follows.

$$\left. \begin{array}{l} a_1 x + b_1 y + c_1 z - l_1 = r_1 \quad (w_1) \\ a_2 x + b_2 y + c_2 z - l_2 = r_2 \quad (w_2) \\ \quad \vdots \quad \quad \vdots \quad \quad \vdots \quad \quad \vdots \quad \quad \vdots \\ a_m x + b_m y + c_m z - l_m = r_m \quad (w_n) \end{array} \right\} \quad (1.35)$$

Note that $l_i = M_i - (a_i X_0 + b_i Y_0 + c_i Z_0)$.

Then, applying the principle of least squares, since

$$[wr^2] = w_1 r_1^2 + w_2 r_2^2 + w_3 r_3^2 + \cdots + w_n r_n^2$$

is to be a minimum, the differential coefficient must be zero and the partial differential coefficients, with respect to each *independent* unknown, must also be zero, i.e.

$$\frac{\partial [wr^2]}{\partial x} = 0, \quad \frac{\partial [wr^2]}{\partial y} = 0, \quad \frac{\partial [wr^2]}{\partial z} = 0$$

Then $\dfrac{\partial [wr^2]}{\partial x} = w_1 r_1 \dfrac{\partial r_1}{\partial x} + w_2 r_2 \dfrac{\partial r_2}{\partial x} + w_3 r_3 \dfrac{\partial r_3}{\partial x} + \ldots$ etc. $= 0$

Taking the general equation

$$w_i r_i^2 = w_i (a_i x + b_i y + c_i z - l_i)^2$$

Then $\dfrac{\partial (w_i r_i^2)}{\partial x} = \dfrac{2 w_i r_i \partial r_i}{\partial x} = 2 w_i a_i (a_i x + b_i y + c_i z - l_i) = 0$

and

$$\frac{\partial [wr^2]}{\partial x} = \begin{pmatrix} w_1 a_1 (a_1 x + b_1 y + c_1 z - l_1) \\ + w_2 a_2 (a_2 x + b_2 y + c_2 z - l_2) \\ \vdots \quad \vdots \quad \vdots \quad \vdots \\ + w_m a_m (a_m x + b_m y + c_m z - l_m) \end{pmatrix} = 0$$

$$= [waa] x + [wab] y + [wac] z - [wal] = 0$$

Applying the same processes to $\partial [wr^2]/\partial y = 0$ and $\partial [wr^2]/\partial z = 0$ gives the following **normal equations**.

$$\begin{aligned} [waa] x + [wab] y + [wac] z - [wal] &= 0 \\ [wba] x + [wbb] y + [wbc] z - [wbl] &= 0 \\ [wca] x + [wcb] y + [wcc] z - [wcl] &= 0 \end{aligned} \quad (1.36)$$

where $[waa] = \Sigma wa^2$; $[wab] = [wba] = \Sigma wab$; etc.

Note that the equations form a matrix with symmetrical coefficients.

Applying matrix algebra to observational equations, the equations may be written as follows.

$$\begin{pmatrix} r_1 = a_{11} x_1 + a_{12} x_2 + \ldots a_{1n} x_n - l_1 \\ r_2 = a_{21} x_1 + a_{22} x_2 + \ldots a_{2n} x_n - l_2 \\ \vdots \\ r_m = a_{m1} x_1 + a_{m2} x_2 + \ldots a_{mn} x_n - l_m \end{pmatrix} \quad (1.37)$$

or in matrix notation

$$r = Ax - l \quad (1.37a)$$

where A is a $(m \times n)$ matrix of known coefficients
x is a $(n \times 1)$ vector of unknowns
l is a $(m \times 1)$ vector of constant terms
r is a $(m \times 1)$ vector of unknown residuals.

1.13 ADJUSTMENT BY OBSERVATIONAL EQUATIONS

$$\underset{m}{\boxed{r}}^{1} = \underset{n}{\boxed{A}}^{1}_{n} \underset{}{\boxed{x}}^{1} - \underset{}{\boxed{l}}^{1} \quad \text{(NB } m \geqslant n\text{)}$$

The most probable values of the unknowns (x) are obtained by the application of least squares, i.e. $[rr] = r^T r = \min$. Then

$$\begin{aligned}(r^T r) &= (Ax - l)^T (Ax - l) = \min \\ &= ((Ax)^T - l^T)(Ax - l) = \min \\ &= (x^T A^T - l^T)(Ax - l) = \min \\ &= (x^T A^T Ax) - (x^T A^T l) - (l^T Ax) + (l^T l) = \min\end{aligned}$$

Partially differentiating and equating to zero gives

$$\frac{\partial (r^T r)}{\partial x} = 2A^T Ax - A^T l - A^T l = 0$$

$$= A^T Ax - A^T l \qquad = 0 \quad \text{(the normal equations)} \tag{1.38}$$

and the solution is

$$x = (A^T A)^{-1} (A^T l) \tag{1.39}$$

If the observations are weighted, i.e. $[wrr] = \min$, the normal equations become

$$(A^T W A x) - (A^T W l) = 0 \tag{1.38a}$$

and the solution is

$$x = (A^T W A)^{-1} (A^T W l) \tag{1.39a}$$

where the weight matrix is

$$W = \begin{pmatrix} w_1 & 0 & \cdots & \cdot \\ 0 & w_2 & & \cdot \\ \vdots & & & \cdot \\ 0 & 0 & \cdots & w_m \end{pmatrix} \tag{1.40}$$

Of particular interest in the matrix solution of x is the matrix $(A^T W A)^{-1}$ from which is derived the **variance–covariance** matrix of the unknowns given by:

$$\sigma_{xx} = \sigma_0^2 (A^T W A)^{-1} = \begin{pmatrix} \sigma_{x_1}^2 & \sigma_{x_1 x_2} & \cdots & \sigma_{x_1 x_n} \\ \sigma_{x_2 x_1} & \sigma_{x_2}^2 & \cdots & \sigma_{x_2 x_n} \\ \vdots & & & \vdots \\ \sigma_{x_n x_1} & \sigma_{x_n x_2} & \cdots & \sigma_{x_n}^2 \end{pmatrix} \tag{1.41}$$

Note that the diagonal terms represent the variances of the unknowns and the remainder, the covariance terms correlating the unknowns. Where there is no correlation the latter terms will be zero.

In many surveying operations it is very useful to have an indication of the precision of the unknowns and the above provides this if a value for σ_0^2 can be found. An unbiased estimate of this value is given as

$$\sigma_0^2 \simeq s_0^2 = \frac{r^T W r}{m - n} \tag{1.42}$$

where m is the number of equations and n is the number of unknowns. If the choice of the weights has been satisfactorily assessed from the standard error of the observations, then σ_0^2 should approach unity.

1.13.1 The error ellipse

Using the variance–covariance matrix, the standard errors of any point in the network are given as σ_E, σ_N and σ_{EN} where σ_E and σ_N represent errors on the axes and σ_{EN} the interaction between them. To find the standard error in an arbitrary direction ϕ, the transformation of coordinates formulae (equations 2.34 and 2.35) are applied, giving

$$E' = E \cos \phi - N \sin \phi \quad \text{and} \quad N' = E \sin \phi + N \cos \phi$$

Thus the standard error ($\sigma_{N'}$) in the N' direction is given by applying the rules for the propagation of variances as

$$\sigma_{E'}^2 = \sigma_N^2 \sin^2 \phi + \sigma_E^2 \cos^2 \phi - 2\sigma_{EN} \sin \phi \cos \phi \tag{1.43}$$

$$\sigma_{N'}^2 = \sigma_N^2 \cos^2 \phi + \sigma_E^2 \sin^2 \phi + 2\sigma_{EN} \sin \phi \cos \phi \tag{1.44}$$

These represent the polar equation of the pedal curve of the error ellipse of the fixation of the point when $\phi = 0$ to 2π (see Fig. 1.4).

To find the maxima and minima values of $\sigma_{N'}$, relative to the variable ϕ, the equation is differentiated and equated to zero as follows.

$$2\sigma_{N'} d(\sigma_{N'}^2)/(d\phi) = 2 \sin \phi \cos \phi \, (\sigma_E^2 - \sigma_N^2)$$
$$+ 2\sigma_{EN} (\cos^2 \phi - \sin^2 \phi) = 0$$

$$2 \tan \phi_m (\sigma_E^2 - \sigma_N^2) = -2 \sigma_{EN} (1 - \tan^2 \phi_m)$$

$$2 \tan \phi_m / (1 - \tan^2 \phi_m) = \tan 2\phi_m = 2\sigma_{EN}/(\sigma_N^2 - \sigma_E^2) \tag{1.45}$$

Substituting the value of ϕ_m back into equation 1.44 gives

$$\sigma_{max} = (\sigma_E^2 + \sigma_N^2)/2 + \sqrt{[(\sigma_E^2 - \sigma_N^2)/4 + \sigma_{EN}^2]} \tag{1.46a}$$

$$\sigma_{min} = (\sigma_E^2 + \sigma_N^2)/2 - \sqrt{[(\sigma_E^2 - \sigma_N^2)/ + \sigma_{EN}^2]} \tag{1.46b}$$

1.13 ADJUSTMENT BY OBSERVATIONAL EQUATIONS   23

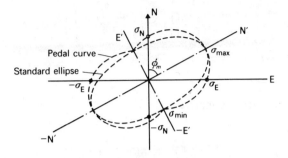

$\sigma_E^2-\sigma_N^2$	σ_{EN}	ϕ_m°
+	−	90-135
−	−	135-180
−	+	180-225
+	+	225-270

Fig. 1.4

Alternatively, if the value of $2\phi_m$ is found by equation 1.45 then

$$\sigma_{max}^2 = (\sigma_E^2 + \sigma_N^2)/2 + \sigma_{EN} \operatorname{cosec} 2\phi_m \qquad (1.47a)$$

$$\sigma_{min}^2 = (\sigma_E^2 + \sigma_N^2)/2 - \sigma_{EN} \operatorname{cosec} 2\phi_m \qquad (1.47b)$$

Applying these principles to solutions involving two unknowns, e.g. intersection where δE and δN are to be found, then the observational equations will be of the form (see p. 55)

$$\begin{pmatrix} r_1 = a_1\delta E + b_1\delta N - l_1 \\ r_2 = a_2\delta E + b_2\delta N - l_2 \\ \vdots \quad \vdots \quad \vdots \quad \vdots \\ r_m = a_m\delta E + b_m\delta N - l_m \end{pmatrix} \begin{array}{l} \text{weight } w_1 \\ \text{weight } w_2 \\ \\ \text{weight } w_m \end{array}$$

i.e. $r = Ax - l \quad (W) \quad$ (see p. 20)

$$(A^T WA) = \begin{pmatrix} [waa] & [wab] \\ [wab] & [wbb] \end{pmatrix}$$

$$(A^T WA)^{-1} = \frac{1}{K}\begin{pmatrix} [wbb] & -[wab] \\ -[wab] & [waa] \end{pmatrix}$$

$$(A^T Wl) = \begin{pmatrix} [wal] \\ [wbl] \end{pmatrix}$$

where K is the determinant $[waa][wbb] - [wab]^2$. The normal equations are then

$$(A^T WA)(x) - (A^T Wl)$$

i.e. $[waa]\delta E + [wab]\delta N - [wal] = 0$
$[wab]\delta E + [wbb]\delta N - [wbl] = 0$

The solution of the equations is, by equation 1.39a,

$$x = (A^T WA)^{-1}(A^T Wl)$$

$$\delta E = (-[wab][wbl] + [wbb][wal])/K \qquad (1.48)$$

$$\delta N = (-[wab][wal] + [waa][wbl])/K \qquad (1.49)$$

i.e. $\begin{pmatrix} \delta E \\ \delta N \end{pmatrix} = \frac{1}{K} \begin{pmatrix} [wbb] \; [wal] & -[wab] \; [wbl] \\ -[wab] \; [wbl] & [waa] \; [wbl] \end{pmatrix}$ (1.50)

The variance is given by

$$s_0^2 \simeq \sigma_0^2 = (r^T W r)/(m-n)$$

The variance–covariance matrix is then

$$\sigma_{xx} = \sigma_0^2 (A^T W A)^{-1}$$

$$= \begin{pmatrix} \sigma_E^2 & \sigma_{EN} \\ \sigma_{EN} & \sigma_N^2 \end{pmatrix} = \frac{\sigma_0^2}{K} \begin{pmatrix} [wbb] & -[wab] \\ -[wab] & [waa] \end{pmatrix}$$

The standard error in the Easting is

$$\sigma_E = \sigma_0 \{[wbb]/K\}^{1/2} \qquad (1.51)$$

The standard error in the Northing is

$$\sigma_N = \sigma_0 \{[waa]/K\}^{1/2} \qquad (1.52)$$

The covariance is

$$\sigma_{EN} = \sigma_0^2 \{-[wab]/K\} \qquad (1.53)$$

Example 1.11 The following observations were made of quantities X, Y and Z. $X = 3.0$ (wt. 1); $X + Y = 6.1$ (wt. 2); $X + Y + Z = 11.2$ (wt. 3); $Y + Z = 8.1$ (wt. 2). Find the most probable values.

Let $X = X_0 + x$ where $X_0 = 3.0$; $Y = Y_0 + y$ where $Y_0 = 3.1$; and $Z = Z_0 + z$ where $Z_0 = 5.1$. The observational equations are then of the form

$$r = ax + by + cz - l \quad \text{weight } w$$

Then $r_1 = 3.0 + x - 3.0$ wt. 1
$r_2 = 3.0 + x + 3.1 + y - 6.1$ wt. 2
$r_3 = 3.0 + x + 3.1 + y + 5.1 + z - 11.2$ wt. 3
$r_4 = \qquad 3.1 + y + 5.1 + z - 8.1$ wt. 2

so $r_1 = x \qquad +0$ wt. 1
$r_2 = x + y \qquad +0$ wt. 2
$r_3 = x + y + z + 0$ wt. 3
$r_4 = \qquad y + z + 0.1$ wt. 2

The normal equations are then of the form

$$[waa]x + [wab]y + [wac]z - [wal] = 0$$
$$[wab]x + [wbb]y + [wbc]z - [wbl] = 0$$
$$[wac]x + [wbc]y + [wcc]z - [wcl] = 0$$

The manual tabulation is then as follows.

1.13 ADJUSTMENT BY OBSERVATIONAL EQUATIONS

a	b	c	−l	Σ	w	waa	wab	wac	−wal	waΣ	wbb	wbc	−wbl	wbΣ	wcc	−wcl	wcΣ
1	0	0	0	1	1	1	0	0	0	1	0	0	0	0	0	0	0
1	1	0	0	2	2	2	2	0	0	4	2	0	0	4	0	0	0
1	1	1	0	3	3	3	3	3	0	9	3	3	0	9	3	0	9
0	1	1	0.1	2.1	2	0	0	0	0	0	2	2	0.2	4.2	2	0.2	4.2
						6	5	3	0	(14)	7	5	0.2	(17.2)	5	0.2	(13.2)

The normal equations are then

$$6x + 5y + 3z + 0 = 0 \quad (14)$$
$$5x + 7y + 5z + 0.2 = 0 \quad (17.2)$$
$$3x + 5y + 5z + 0.2 = 0 \quad (13.2)$$

Solving these equations simultaneously gives $x = 0.055$, $y = -0.055$ and $z = -0.016$.

Then $X = 3.0 + 0.055 = 3.055$
$Y = 3.1 - 0.055 = 3.045$
$Z = 5.1 - 0.016 = 5.084$

Solving the problem by matrix methods, $r = Ax - l \quad (W)$,

$$r = \begin{pmatrix} 1 & 0 & 0 \\ 1 & 1 & 0 \\ 1 & 1 & 1 \\ 0 & 1 & 1 \end{pmatrix} \begin{pmatrix} X \\ Y \\ Z \end{pmatrix} - \begin{pmatrix} 3.0 \\ 6.1 \\ 11.2 \\ 8.1 \end{pmatrix} \quad \begin{pmatrix} 1 & 0 & 0 & 0 \\ 0 & 2 & 0 & 0 \\ 0 & 0 & 3 & 0 \\ 0 & 0 & 0 & 2 \end{pmatrix}$$

$$\quad\quad A \quad\quad\quad x \quad\quad\quad l \quad\quad\quad\quad\quad W$$

The normal equations are given as $(A^T W A x) - (A^T W l)$, i.e.

$$\begin{pmatrix} 1 & 0 & 0 & 0 \\ 0 & 2 & 0 & 0 \\ 0 & 0 & 3 & 0 \\ 0 & 0 & 0 & 2 \end{pmatrix} \quad \begin{pmatrix} 1 & 0 & 0 \\ 1 & 1 & 0 \\ 1 & 1 & 1 \\ 0 & 1 & 1 \end{pmatrix} \quad \begin{pmatrix} 3.0 \\ 6.1 \\ 11.2 \\ 8.1 \end{pmatrix}$$

$$\begin{pmatrix} 1 & 1 & 1 & 0 \\ 0 & 1 & 1 & 1 \\ 0 & 0 & 1 & 1 \end{pmatrix} \begin{bmatrix} 1 & 2 & 3 & 0 \\ 0 & 2 & 3 & 2 \\ 0 & 0 & 3 & 2 \end{bmatrix} \begin{bmatrix} 6 & 5 & 3 \\ 5 & 7 & 5 \\ 3 & 5 & 5 \end{bmatrix} \quad \begin{pmatrix} 1 & 2 & 3 & 0 \\ 0 & 2 & 3 & 2 \\ 0 & 0 & 3 & 2 \end{pmatrix} \begin{bmatrix} 48.8 \\ 62.0 \\ 49.8 \end{bmatrix}$$

$$\quad\quad\quad\quad A^T W \quad\quad A^T W A \quad\quad\quad\quad\quad\quad A^T W l$$

The normal equations are then

$$6X + 5Y + 3Z - 48.0 = 0$$
$$5X + 7Y + 5Z - 62.0 = 0$$
$$3X + 5Y + 5Z - 49.8 = 0$$

The solution to the normal equation is

$$x = (A^T W A)^{-1} (A^T W l)$$

But

$$(A^T W A)^{-1} = \frac{1}{22}\begin{pmatrix} 10 & -10 & 4 \\ -10 & 21 & -15 \\ 4 & -15 & 17 \end{pmatrix}$$

$$A^T W l \begin{pmatrix} 48.8 \\ 62.0 \\ 49.8 \end{pmatrix}$$

$$\begin{pmatrix} 0.4545 & -0.4545 & 0.1818 \\ -0.4545 & 0.9545 & -0.6818 \\ 0.1818 & -0.6818 & 0.7727 \end{pmatrix} \begin{bmatrix} 3.054 \\ 3.046 \\ 5.082 \end{bmatrix}$$

$(A^T W A)^{-1}$ $\qquad\qquad x$

Then $X = 3.054$, $Y = 3.046$, $Z = 5.082$.

1.14 Application to levelling circuits

1) Draw a circuit diagram and indicate the direction of the levelling.
2) If no lengths are given or a weight is not attributed, then the errors are assumed equal.
3) Where the lengths of the lines are given and all but the compensating errors have been eliminated, the number of settings of the level is assumed proportional to the length (S). Then

$$s_x \propto \sqrt{n} \propto \sqrt{S}$$

where n is the number of sets and S is the length of the line. Thus

$$w \propto 1/s_x^2 \propto 1/S \qquad (1.54)$$

i.e. the weight is inversely proportional to the length of the line.

Note that the acceptable closing error in a circuit may be written as $x\sqrt{n}$ mm where x may be 1, 2 or 3 according to the type of work in hand.

4) If a line is levelled N times then the mean value has a standard error of $s_{\bar{x}} = s_x/\sqrt{N}$. Therefore

$$s_{\bar{x}} \propto 1/\sqrt{N}$$

and $\quad w \propto 1/(1/\sqrt{N})^2 \propto N \qquad (1.55)$

Example 1.12 In a circuit of levels $ABCDA$ (see Fig. 1.5) the following results were obtained: $A-B+5.216$, $B-C+2.394$, $C-D+1.055$, $D-A-8.690$. Calculate the most probable height differences assuming the following alternative conditions.

(a) All the observations are of equal accuracy; (b) the lines BC and CD are twice as long as the lines AB and DA; (c) the lines BC and CD have been levelled twice and the values given are the mean values.

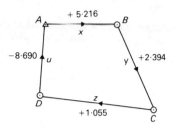

Fig. 1.5

Let the level differences be

$AB = X = 5.216$
$BC = Y = 2.394$
$CD = Z = 1.055$
$DA = U = -8.690$ (this is assumed dependent).

Assuming

$X = X_0 + x = 5.216 + x$
$Y = Y_0 + y = 2.394 + y$
$Z = Z_0 + z = 1.055 + z$

(a) The observational equations become

$r_1 = x \quad\quad +0$
$r_2 = \quad\quad y \quad +0$
$r_3 = \quad\quad\quad\quad z+0$
$r_4 = x+y+z-0.025$

By inspection the normal equations become

$2x + y + z - 0.025 = 0$
$x + 2y + z - 0.025 = 0$
$x + y + 2z - 0.025 = 0$

Solving these equations gives $x = y = z = 0.025/4 = 0.0063$. Then $X = 5.222(3)$, $Y = 2.400(3)$, $Z = 1.061(3)$ and $U = -8.683(9)$.

(b) Here $w \propto 1/S$, i.e.

$w_x : w_y : w_z :: 1/1 : 1/2 : 1/2 : 1/1 :: 2 : 1 : 1 : 2$

The observational equations now become

$r_1 = x \quad\quad +0 \quad\quad$ wt. 2
$r_2 = \quad\quad y \quad +0 \quad\quad$ wt. 1
$r_3 = \quad\quad\quad\quad z+0 \quad\quad$ wt. 1
$r_4 = x+y+z-0.025 \quad\quad$ wt. 2

To derive the normal equations multiply by the weights and then equate to zero, i.e.

$2x \quad\quad\quad\quad\quad\quad = 0$
$\quad\quad y \quad\quad\quad\quad = 0$
$\quad\quad\quad\quad z \quad\quad = 0$
$2x + 2y + 2z - 0.050 = 0$

By inspection the normal equations are then

$$4x + 2y + 2z - 0.050 = 0$$
$$2x + 3y + 2z - 0.050 = 0$$
$$2x + 2y + 3z - 0.050 = 0$$

Solving these equations gives $x = 0.004(2)$, $y = 0.008(3)$ and $z = 0.008(3)$. Then $X = 5.220(2)$, $Y = 2.402(3)$, $Z = 1.063(3)$ and $U = -8.685(8)$.

(c) Here the number of observations are different, i.e. $w \propto N$ and

$$w_x : w_y : w_z : w_u : : 1 : 2 : 2 : 1$$

The observational equations now become

$$r_1 = x \qquad + 0 \qquad \text{wt. 1}$$
$$r_2 = \quad y \quad + 0 \qquad \text{wt. 2}$$
$$r_3 = \qquad\quad z + 0 \qquad \text{wt. 2}$$
$$r_4 = x + y + z - 0.025 \quad \text{wt. 1}$$

Multiplying by the weight and equating to zero as before gives

$$x \qquad\qquad\qquad = 0$$
$$\quad 2y \qquad\qquad = 0$$
$$\qquad\quad 2z \qquad = 0$$
$$x + y + z - 0.025 = 0$$

By inspection the normal equations are then

$$2x + y + z - 0.025 = 0$$
$$x + 3y + z - 0.025 = 0$$
$$x + y + 3z - 0.025 = 0$$

Solving these equations gives $x = 0.008(3)$, $y = 0.004(2)$ and $z = 0.004(2)$. Then $X = 5.224(3)$, $Y = 2.398(2)$ and $Z = 1.059(2)$.

1.15 Station adjustment

This is only needed when the angles or directions have been observed in various combinations. During triangulation, angular observations to all the stations in the round may not be possible and thus some angles may be measured more times than others.

Note that

$$s_{\bar{x}} \propto s_x/\sqrt{n} \quad \text{and} \quad w \propto 1/s_{\bar{x}}^2$$
$$s_{\bar{x}} \propto 1/\sqrt{n} \qquad\qquad\qquad\qquad (1.56)$$

Therefore $w \propto n$, i.e. the weight of the mean value is proportional to the number of observations assuming equal reliability.

1.15 STATION ADJUSTMENT

Example 1.13 The following table of mean values of angles was obtained from observations taken at a triangulation station A and the number of times the respective angles were repeated.

Angle	Mean value	Number of repetitions
BAE	146° 27′ 31.2″	10
BAC	35° 17′ 48.6″	6
CAD	64° 45′ 31.0″	8
DAE	46° 24′ 06.4″	6

Compute the most probable value of the angles.

Let $BAC = 35° 17′ 48.6″ + a$ wt. 3
$CAD = 64° 45′ 31.0″ + b$ wt. 4
$DAE = 46° 24′ 06.4″ + c$ wt. 3

The observational equations become

$35° 17′ 48.6″ + a + 64° 45′ 31.0 + b + 46° 24′ 06.4″ + c - 146° 27′ 31.2″ = r_1$ (wt. 5)
$35° 17′ 48.6 + a \qquad\qquad\qquad - 35° 17′ 48.6″ = r_2$ (wt. 3)
$\qquad\qquad 64° 45′ 31.0 + b \qquad - 64° 45′ 31.0″ = r_3$ (wt. 4)
$\qquad\qquad\qquad\qquad 46° 24′ 06.4″ + c - 46° 24′ 06.4″ = r_4$ (wt. 3)

i.e. $a + b + c - 5.2″ = r_1$ wt. 5
$\qquad a \qquad\qquad = r_2$ wt. 3
$\qquad\quad b \qquad = r_3$ wt. 4
$\qquad\qquad c = r_4$ wt. 3

These results may be tabulated as follows.

a	b	c	−1	Σ	w	waa	wab	wac	−waΣ	wa	wbb	wbc	−wbΣ	wb	wcc	−wcl	wcΣ
1	1	1	−5.2	−2.2	5	5	5	5	−26.0	−11.0	5	5	−26.0	−11.0	5	−26.0	−11.0
1			0		1.0	3	3			3.0							
	1		0		1.0	4					4			4.0			
		1	0		1.0	3									3		3.0
2	2	2	−5.2	0.8	8	5	5	−26.0	−8.0	9	5	−26.0	−7.0	8	−26.0	−8.0

The normal equations are then

$8a + 5b + 5c - 26.0 = 0$
$5a + 9b + 5c - 26.0 = 0$
$5a + 5b + 8c - 26.0 = 0$

Solving these equations simultaneously gives

$a = c = 1.55″$ and $b = 1.16″$

The most probable values of the angles are then

$BAC = 35° 17′ 48.6 + 1.55″ = 35° 17′ 50.15″$ (50.2″)

$$CAD = 64°\,45'\,31.0'' + 1.16'' = 64°\,45'\,32.16'' \quad (32.2'')$$
$$DAE = 46°\,24'\,06.4'' + 1.55'' = 46°\,24'\,07.99'' \quad (08.0'')$$

Note that the latter two examples are simple adjustment problems which illustrate the principles of observational equations. As a more simple solution, the errors may be distributed as inversely proportional to the weights. Thus the error of 5.2" is distributed in the ratio

$$1/5:1/3:1/4:1/3$$
i.e. $\quad 12:20:15:20 \quad (67)$

Then $\quad e_1 = 12/67 \times 5.2 = 0.931''$
$ e_2 = 20/67 \times 5.2 = 1.552''$
$ e_3 = 15/67 \times 5.2 = 1.164''$
$ e_4 = e_2 = \underline{1.552''}$
$ \text{Check} = 5.199''$

Example 1.14 The following results were obtained in observations made to determine the instrument constants of a theodolite to be used for tacheometry.

Given that the horizontal length is given as

$$D = ms\cos^2\theta + K\cos\theta$$

and that the telescope was level throughout, compute the most probable value of m and K.

Distances (D) (metres)   20.00   40.00   60.00   80.00   100.00
Staff intercept (s)   0.398   0.798   1.199   1.599   1.998

(RICS/M)

As the telescope is level, $\theta = 0$ and therefore the general observational equation may be written of the form

$$ms_i + K - D_i = r_i$$

i.e.
$$0.398m + K - 20.00 = r_1$$
$$0.798m + K - 40.00 = r_2$$
$$1.199m + K - 60.00 = r_3$$
$$1.599m + K - 80.00 = r_4$$
$$1.998m + K - 100.00 = r_5$$

The results may be tabulated as follows.

a	b	−M	Σ	aa	ab	−aM	aΣ	bb	−bM	bΣ
0.398	1	−20.00	−18.602	0.158	0.398	−7.960	−7.404	1	−20.000	−18.602
0.798	1	−40.00	−38.202	0.637	0.798	−31.920	−30.485	1	−40.000	−38.202
1.199	1	−60.00	−57.801	1.438	1.199	−71.940	−69.303	1	−60.000	−57.801
1.599	1	−80.00	−77.401	2.557	1.599	−127.920	−123.764	1	−80.000	−77.401
1.998	1	−100.00	−97.002	3.992	1.998	−199.800	−193.810	1	−100.000	−97.002
				8.782	5.992	−439.540	−424.766	5	−300.000	−289.008

The normal equations are then

$$8.782m + 5.992K - 439.540 = 0$$
$$5.992m + 5.000K - 300.000 = 0$$

i.e.
$$m + 0.682K - 50.050 = 0$$
$$m + 0.834K - 50.067 = 0$$

Then
$$-0.152K + 0.017 = 0$$

Therefore

$$K = 0.112 \quad (0.1)$$
$$m = 49.974 \quad (50)$$

The practical values are then $m = 50$ and $K = 0.1$.

Exercises 1.2

1 A straight line $ABCD$ was measured as a whole and in sectons. Due to variations in the accuracy of the measurements, weights have been assigned as shown below.

	Measured length (m)	Weight
AB	39.231	3
BC	120.716	2
CD	61.256	2
AC	159.935	1
AD	221.218	1

Find the most probable lengths of AB, BC and CD to the nearest 0.0001 m. (LU Ans. 39.2307 m, 120.7156 m, 61.2612 m)

2 As part of a triangulation the following observations were made.

A	62° 02′ 41″
B	44° 21′ 15″
C	87° 12′ 31″
$A+B$	106° 23′ 54″
$B+C$	131° 33′ 49″

Calculate the most probable values of the angles A, B and C.
 (RICS/M Ans. 62° 02′ 39.9″, 44° 21′ 15.3″, 87° 12′ 32.4″)

3 Levels have been taken on to an existing roadway thought to be a vertical curve of the form $y = ax^2 + bx + c$. Given the recorded data as follows, find the most probable value of a, b and c.

	Chainages	10.00 m	20.00 m	30.00 m	40.00 m
	Levels	63.35 m	63.00 m	62.88 m	62.97 m

(TP Ans. $a = 0.0011$, $b = -0.0676$, $c = 63.915$ m)

4 The following coordinates refer to the position of a number of theodolite stations along an underground roadway.

	E (m)	N (m)
A	457 040.57	347 467.45
B	457 090.79	347 513.52
C	457 197.50	347 628.00
D	457 331.62	347 775.86
X	456 797.00	347 206.00
Y	457 498.30	347 950.20

Calculate (a) the offsets from A, B, C and D on to the line $X\,Y$; (b) the offsets from A, B, C, D, X and Y on to the best fit line. (TP Ans. (a) 2.05 m, -2.91 m, -2.06 m, 1.74 m. (b) 2.32 m, -2.66 m, -1.88 m, 1.82 m, 0.45 m, 0.03 m)

1.16 Adjustment by conditional equations (direct method)

Whenever there is an observational or geometrical condition to be fulfilled, the most probable values of the measured quantities may be derived by the use of disappearing multipliers, known as **correlatives**.

As before, each observed quantity (M_i) is subject to a residual error (r_i), i.e.

$$X_i = M_i + r_i$$

The conditional equations will then be of the form

$$A = \begin{pmatrix} a_1 r_1 + a_2 r_2 + \ldots + a_n r_n = Q_1 - [aM] = Y_1 \\ b_1 r_1 + b_2 r_2 + \ldots + b_n r_n = Q_2 - [bM] = Y_2 \\ \vdots \quad \vdots \quad \vdots \quad \vdots \quad \vdots \quad \vdots \quad \vdots \\ m_1 r_1 + m_2 r_2 + \ldots + m_n r_n = Q_m - [mM] = Y_m \end{pmatrix}$$

where $m < n$ (1.57)

Consider now the application of the least square condition together with the conditional equations as a function F as follows.

$$F = w_1 r_1^2 + w_2 r_2^2 + \ldots + w_n r_n^2$$
$$- 2k_1 (a_1 r_1 + a_2 r_2 + \ldots + a_n r_n - Y_1)$$
$$- 2k_2 (b_1 r_1 + b_2 r_2 + \ldots + b_n r_n - Y_2)$$
$$\vdots$$
$$- 2k_m (m_1 r_1 + m_2 r_2 + \ldots + m_n r_n - Y_n)$$

The coefficients k_i are the correlatives and it can be seen that their number equals the number of conditions.

To find the most probable values for the residual errors, the function F is partially differentiated with respect to each of the variables and equated to zero, i.e.

$$\partial F/\partial r_1 = 2w_1 r_1 - 2k_1 a_1 - 2k_2 b_1 - \ldots - 2k_m m_1 = 0$$

and thus the residuals may be written as

$$B = \begin{pmatrix} r_1 = (1/w_1)(k_1 a_1 + k_2 b_1 + \ldots + k_m m_1) \\ r_2 = (1/w_2)(k_1 a_2 + k_2 b_2 + \ldots + k_m m_2) \\ \vdots \quad \vdots \quad \vdots \quad \quad \vdots \\ r_n = (1/w_n)(k_1 a_n + k_2 b_n + \ldots + k_m m_n) \end{pmatrix} \quad (1.58)$$

Substituting the values of the residuals in (B) into the conditional equations (A) gives

$$C = \begin{pmatrix} a_1 r_1 = (a_1/w_1)(k_1 a_1 + k_2 b_1 + \ldots + k_m m_1) \\ a_2 r_2 = (a_2/w_2)(k_1 a_2 + k_2 b_2 + \ldots + k_m m_2) \\ \vdots \quad \vdots \quad \quad \vdots \\ a_n r_n = (a_n/w_n)(k_1 a_n + k_2 b_n + \ldots + k_m m_n) \end{pmatrix} \quad (1.59)$$

Taking all of the conditional equations together gives

$$\begin{pmatrix} [aa/w]k_1 + [ab/w]k_2 + \ldots + [am/w]k_m = Y_1 \\ [ba/w]k_1 + [bb/w]k_2 + \ldots + [bm/w]k_m = Y_2 \\ \vdots \quad \quad \quad \vdots \\ [ma/w]k_1 + [mb/w]k_2 + \ldots + [mm/w]k_m = Y_m \end{pmatrix} \quad (1.60)$$

These equations represent the **correlative normal equations** and the coefficients are symmetrical as before.

The normal equations are now solved for $k_1, k_2, \ldots, k_m$ and then substituted back into the equations (B) to obtain the values of the residuals.

The matrix notation is similar to that for observational equations, i.e.

$$Ar = Y \quad (W) \quad \text{and} \quad r = W^{-1} A^T k \quad (1.61)$$

where A is a $(m \times n)$ matrix of known coefficients
 r is a $(n \times 1)$ vector of unknowns
 Y is a $(m \times 1)$ vector of constant terms.

Simultaneous normal (linear) equations for the unknown correlatives (k) with weights (w) applied to the residuals (r) are

$$(AW^{-1}A^T)(k) = (Y) \quad (1.62)$$

The solution for the correlatives is

$$(k) = (AW^{-1}A^T)(Y) \quad (1.63)$$

Back substitution into the conditional equations gives the most probable values of the residuals as

$$(r) = (w^{-1}A^T)(Aw^{-1}A^T)^{-1}(Y) \qquad (1.64)$$

Example 1.15 Given that $x_1 = 3.00$ (wt. 2), $x_2 = 5.00$ (wt. 1), $x_3 = 7.00$ (wt. 1), $x_4 = 2.63$ (wt. 2), $x_5 = 3.27$ (wt. 1), subject to the conditions $x_1 + x_2 + x_3 + x_4 = 17.66$, and $x_3 + x_4 + x_5 = 12.96$, find the most probable values x_1 to x_5.

Let the measured values of x be subject to residual errors; r_1, r_2, r_3, r_4, r_5. Then the conditional equations become

$3.00 + r_1 + 5.00 + r_2 + 7.00 + r_3 + 2.63 + r_4 = 17.66$

i.e. $\quad r_1 \quad\quad + r_2 \quad\quad + r_3 \quad\quad + r_4 = 17.66 - 17.63$

$$= 0.03 \qquad (1)$$

and similarly

$r_3 \quad\quad + r_4 \quad\quad + r_5 \quad\quad = 12.96 - 12.90$

$$= 0.06 \qquad (2)$$

The least squares condition is that the expression

$$w_1 r_1^2 + w_2 r_2^2 + w_3 r_3^2 + w_4 r_4^2 + w_5 r_5^2$$

shall be a minimum. As suggested by Jameson the partial differentiation gives, in tabular form,

$$w_1 r_1 \delta r_1 + w_2 r_2 \delta r_2 + w_3 r_3 \delta r_3 + w_4 r_4 \delta r_4 + w_5 r_5 \delta r_5 = 0$$
$$\delta r_1 + \quad \delta r_2 + \quad \delta r_3 + \quad \delta r_4 + \quad \delta r_5 = 0 \times (-k_1)$$
$$\delta r_3 + \quad \delta r_4 + \quad \delta r_5 = 0 \times (-k_2)$$

Then $\quad \delta r_1(w_1 r_1 - k_1) + \delta r_2(w_2 r_2 - k_2) + \delta r_3(w_3 r_3 - k_1 - k_2)$
$$+ \delta r_4(w_4 r_4 - k_1 - k_2) + \delta r_5(w_5 r_5 - k_2) = 0$$

But each δr is independent, therefore each term individually must disappear, i.e.

$$\begin{array}{ll} w_1 r_1 - k_1 = 0 & \text{or} \quad r_1 = k_1/w_1 \\ w_2 r_2 - k_1 = 0 & r_2 = k_1/w_2 \\ w_3 r_3 - k_1 - k_2 = 0 & r_3 = (k_1 + k_2)/w_3 \\ w_4 r_4 - k_1 - k_2 = 0 & r_4 = (k_1 + k_2)/w_4 \\ w_5 r_5 - k_2 = 0 & r_5 = k_2/w_5 \end{array}$$

Substituting these values back into the conditional equations,

(1) $k_1/w_1 + k_1/w_2 + (k_1 + k_2)/w_3 + (k_1 + k_2)/w_4 = 0.03$
(2) $(k_1 + k_2)/w_3 + (k_1 + k_2)/w_4 + k_2/w_5 \quad = 0.06$

i.e. $6k_1 + 3k_2 = 0.06$

$3k_1 + 5k_2 = 0.12$

Solving simultaneously gives $k_1 = -0.003$ and $k_2 = 0.026$. The residual values are then

$$r_1 = -0.003/2 = -0.0015$$

1.16 ADJUSTMENT BY CONDITIONAL EQUATIONS 35

$$r_2 = -0.003/1 = -0.0030$$
$$r_3 = (-0.003 + 0.026)/1 = 0.0230$$
$$r_4 = (-0.003 + 0.026)/2 = 0.0115$$
$$r_5 = 0.026/1 = 0.0260$$

The most probable values then become

$$\left.\begin{array}{l} x_1 = 3.0000 - 0.0015 = 2.9985 \\ x_2 = 5.0000 - 0.0030 = 4.9970 \\ x_3 = 7.0000 + 0.0230 = 7.0230 \\ x_4 = 2.6300 + 0.0115 = 2.6415 \\ x_5 = 3.2700 + 0.0260 = 3.2960 \end{array}\right\} \begin{array}{l} 17.66(00) \\ \\ 12.96(05) \end{array}$$

Considering equation 1.60 the equations may be tabulated as follows.

	ak_1	bk_2	Σ	$1/w$	aa/w	ab/w	$a\Sigma/w$	bb/w	$b\Sigma/w$
r_1	1	0	1	0.5	0.5	0	0.5	0	0
r_2	1	0	1	1.0	1.0	0	1.0	0	0
r_3	1	1	2	1.0	1.0	1.0	2.0	1.0	2.0
r_4	1	1	2	0.5	0.5	0.5	1.0	0.5	1.0
r_5	0	1	1	1.0	0	0	0	1.0	1.0
	4	3	7		3.0	1.5	4.5	2.5	4.0

The correlative normal equations are then as before

$$3.0k_1 + 1.5k_2 = 0.03$$
$$1.5k_1 + 2.5k_2 = 0.06$$

In the table above, the observational coefficients are shown vertically; the check column helps the arithmetic; the alternative headings of the columns k_1, k_2 etc. give the ultimate residual equations, e.g. $r_3 = (1k_1 + 1k_2) \times 1/w_3$ (equation 1.58); and the summation values provide the coefficients of the normal equations. These provide the minimum number and the symmetry of the equations allows the missing values to be inserted.

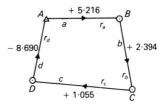

Fig. 1.6

Example 1.16 In a circuit of levels, $ABCDA$, the following results were obtained; AB +5.216 m, BC +2.394 m, CD +1.055 m, DA −8.690 m (see Fig. 1.6). Calculate the most probable heights of B, C and D above A assuming the alternative conditions (a) all the observations are of equal accuracy; (b) the lines BC and CD are twice as long as the lines AB and DA; (c) the lines BC and CD have been levelled twice and the values given are the mean values.

(a) Adopting the conditional equations solution, let the level differences (Δh) be as follows.

$$AB = a = 5.216 + r_a$$
$$BC = b = 2.394 + r_b$$
$$CD = c = 1.055 + r_c$$
$$DA = d = -8.690 + r_d$$

The conditional equation is then

$$r_a + r_b + r_c + r_d + 5.216 + 2.394 + 1.055 - 8.690 = 0$$

As the observations are of equal accuracy, the correlative k is given by

$$k = 0.025/4 = 0.00625$$

Then $r_a = r_b = r_c = r_d = 0.00625$

$$a = 5.216 + 0.006(3) = 5.222(3)$$
$$b = 2.394 + 0.006(3) = 2.400(3)$$
$$c = 1.055 + 0.006(3) = 1.061(3)$$
$$\overline{}$$
$$8.683(9) = 8.684$$

$$d = -8.690 + 0.006(3) = -8.683(7) = 8.684$$
(Check)

The levels are then given relative to $A = 0$ as $B = +5.222$ m, $C = +7.622$ m, $D = 8.684$ m.

(b) Here the lengths (S) are different and the weights $\propto 1/S$, i.e.

$$w_a : w_b : w_c : w_d$$
$$1/1 : 1/2 : 1/2 : 1/1 \text{ or } 2:1:1:2$$

Substituting these values into the conditional equation,

$$k/2 + k/1 + k/1 + k/2 = 0.025$$
$$k = 0.025/3 = 0.0083$$

Then $r_a = 0.0083/2 = 0.004(2)$
$r_b = 0.0083/1 = 0.008(3)$
$r_c = 0.0083/1 = 0.008(3)$
$r_d = 0.0083/2 = 0.004(2)$

Then $a = 5.216 + 0.004(2) = 5.220(2)$
$b = 2.394 + 0.008(3) = 2.402(3)$
$c = 1.055 + 0.008(3) = 1.063(3)$
$$\overline{}$$
$$8.685(8) = 8.686 \text{ m}$$

$$d = -8.690 + 0.004(2) = -8.685(8) = 8.686 \text{ m}$$
(Check)

(c) Here the number of observations are different: $w \propto n$, and

$$w_a : w_b : w_c : w_d$$
$$1 : 2 : 2 : 1$$

1.16 ADJUSTMENT BY CONDITIONAL EQUATIONS

Then $k + k/2 + k/2 + k = 0.025$

$k = 0.025/3 = 0.0083$

but $r_a = 0.0083/1 = 0.008(3)$
$r_b = 0.0083/2 = 0.004(2)$
$r_c = 0.0083/2 = 0.004(2)$
$r_d = 0.0083/1 = 0.008(3)$

Then $a = 5.216 + 0.008(3) = 5.224(3)$
$b = 2.394 + 0.004(2) = 2.398(2)$
$c = 1.055 + 0.004(2) = 1.059(2)$

$$8.681(7) = 8.682 \text{ m}$$

$d = -8.690 + 0.008(3) = 8.681(7) = 8.682 \text{ m}$ (Check)

Example 1.17 In an EDM traverse network, the following partial coordinates have been computed.

	ΔE	ΔN
AB	1105.362	1346.542
BC	964.547	−965.426
CD	−892.513	−882.492
DA	−1177.341	501.334
BD	72.084	−1847.982

Assuming all the measurements were of equal accuracy, compute the most probable coordinates of stations B, C and D, given that the coordinates of A are 1000.000 mE, 1000.000 m N.
(RICS/M)

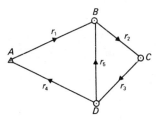

Fig. 1.7

Let the partial Eastings be subject to the residual errors as shown in Fig. 1.7. The conditional equations may be written as

$r_1 + r_2 + r_3 + r_4 \quad + 0.055 = 0$
$\quad r_2 + r_3 \quad + r_5 - 0.050 = 0$

where $\Sigma \Delta E$ for the circuit $ABCDA = 0.055 \text{ m}$
$\Sigma \Delta E$ for the circuit $BCDB = -0.050 \text{ m}$.

Tabulating for the correlative normal equations,

	a\k₁	b\k₂	Σ	aa	ab	aΣ	bb	bΣ
r_1	1	0	1	1	0	1	0	0
r_2	1	1	2	1	1	2	1	2
r_3	1	1	2	1	1	2	1	2
r_4	1	0	1	1	0	1	0	0
r_5	0	1	1	0	0	0	1	1
				4	2	6	3	5

38 ERRORS AND THEIR ADJUSTMENT

The normal equations are then

$$4k_1 + 2k_2 + 0.055 = 0$$
$$2k_1 + 3k_2 - 0.050 = 0$$

Solving gives $k_1 = -0.0331$ and $k_2 = 0.0388$.

The Northings are dealt with in a similar manner and the coefficients of the correlatives will be the same but the absolute values are different, i.e.

$$4k'_1 + 2k'_2 - 0.042 = 0$$
$$2k'_1 + 3k'_2 + 0.064 = 0$$

Solving gives $k'_1 = 0.0318$ and $k'_2 = -0.0425$.

The residuals are then as follows.

	ΔE		ΔN	
$r_1 = k_1$	$= -0.0331$		0.0318	
$r_2 = k_1 + k_2$	$= 0.0057$	$-0.054\,(8)$	-0.0107	$0.042\,(2)$
$r_3 = k_1 + k_2$	$= 0.0057$		-0.0107	
$r_4 = k_1$	$= -0.0331$		0.0318	
$r_5 = k_2$	$= 0.0388$	$0.050\,(2)-0.0425$		$-0.063\,(9)$

The total coordinates are then as follows.

		E (m)			N (m)
A	ΔE	1000.0000		ΔN	1000.0000
	1105.3620 − 0.0331 =	1105.3289		1346.5420 + 0.0318 =	1346.5738
B		2105.3289			2346.5738
	964.5470 + 0.0057 =	964.5527		− 965.4260 − 0.0107 =	− 965.4367
C		3069.8816			1381.1371
	− 892.5130 + 0.0057 =	− 892.5073		− 882.4920 − 0.0107 =	− 882.5027
D		2177.3743			498.6344
	− 1177.3410 − 0.0331 =	− 1177.3741		501.3340 + 0.0318 =	501.3658
A		1000.0002			1000.0002

The coordinates are then

B   2105.329 m E   2346.574 m N
C   3069.882 m E   1381.137 m N
D   2177.374 m E   498.634 m N

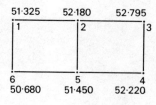

Fig. 1.8

Example 1.18 Figure 1.8 shows part of a concrete slab which has to be screeded to form a regularly inclined plane. Assuming that all the levels shown, which are part of a 100 m grid, are of equal accuracy, complete the most probable values of the formation levels given that the minimum thickness of screed is to be 20 mm.

1.16 ADJUSTMENT BY CONDITIONAL EQUATIONS

The conditions to be fulfilled are as follows:

i) $1-2 = 2-3$ i.e. $1-2(2)+3 = 0$
ii) $6-5 = 5-4$ i.e. $6-2(5)+4 = 0$
iii) $1-6-2+5 = 0$
iv) $2-5-3+4 = 0$

As (i) – (ii) = (iii) – (iv) then only three equations are independent. Equation (iv) is rejected.

Let

$1 = 51.325 + r_1 \qquad 2 = 52.180 + r_2 \qquad 3 = 52.795 + r_3$
$4 = 52.220 + r_4 \qquad 5 = 51.450 + r_5 \qquad 6 = 50.680 + r_6$

The conditional equations then become

$$r_1 - 2r_2 + r_3 \qquad\qquad -0.240 = 0$$
$$r_4 - 2r_5 + r_6 \qquad\qquad = 0$$
$$r_1 - r_2 \qquad + r_5 - r_6 - 0.085 = 0$$

The results may be tabulated as follows.

	$a\backslash k_1$	$b\backslash k_2$	$c\backslash k_3$	Σ	aa	ab	ac	aΣ	bb	bc	bΣ	cc	cΣ
r_1	1	0	1	2	1	0	1	2	0	0	0	1	2
r_2	-2	0	-1	-3	4	0	2	6	0	0	0	1	3
r_3	1	0	0	1	1	0	0	1	0	0	0	0	0
r_4	0	1	0	1	0	0	0	0	1	0	1	0	0
r_5	0	-2	1	-1	0	0	0	0	4	-2	2	1	-1
r_6	0	1	-1	0	0	0	0	0	1	-1	0	1	0
					6	0	3	9	6	-3	3	4	4

The correlative normal equations are then

$$6k_1 \qquad + 3k_3 - 0.240 = 0$$
$$6k_2 - 3k_3 \qquad = 0$$
$$3k_1 - 3k_2 + 4k_3 - 0.085 = 0$$

Solving gives $k_1 = 0.0575$, $k_2 = -0.0175$ and $k_3 = -0.0350$.

Thus $r_1 = 0.0225 \qquad r_2 = -0.0800 \qquad r_3 = 0.0575$
$r_4 = -0.0175 \qquad r_5 = 0.0 \qquad r_6 = 0.0175$.

	Existing	r	Plane	Formation	Δmm
At 1	51.3250	0.0225	51.3475	51.448	+123
2	52.1800	-0.0800	52.1000	52.200	+ 20
3	52.7950	0.0575	52.8525	52.953	+158
4	52.2200	-0.0175	52.2025	52.303	+ 83
5	51.4500	0.0	51.4500	51.550	+100
6	50.6800	0.0175	50.6975	50.798	+118

Note that the highest point above the plane of formation is 2, i.e. 80 mm above. Therefore the minimum screed is applied here, 20 mm above the existing ground and thus lifting the formation plane 100 mm above the theoretical plane.

Exercises 1.3

1 In order to establish the value of three bench marks B, C and D, two levelling circuits were formed, $ABCDA$ and $ACDA$; the following results were recorded.

Circuit 1: AB, $+3.753$ m; BC, $+5.548$ m; CD, $+10.427$ m; DA, -19.721 m.
Circuit 2: AC, $+9.280$ m; CB, -5.540 m; BA, -3.755 m.

Calculate the most probable value of the bench marks if the reduced level of A is 169.721 m AOD.

(Ans. B, 173.471 m; C, 179.011 m; D, 189.440 m)

2 A and C are two bench marks where C is known to be 26.502 m above A. Given the circuit data below, find the most probable values of the levels of B and D relative to A.

AB	3 km	$+11.460$ m
BC	4 km	$+15.130$ m
CD	4 km	-9.339 m
DB	6 km	-5.703 m
DA	3 km	-17.111 m

(Ans. $+11.424$ m, $+17.132$ m)

3 Undernoted are the results of two rounds of precise trigonometrical levellings between four bench marks.

AB	2.25 km	$+39.274$ m	BA	-39.243 m
BC	4.00 km	-94.848 m	CB	$+94.892$ m
CD	1.75 km	$+20.052$ m	DC	-20.032 m
DA	3.00 km	$+35.619$ m	AD	-35.587 m

Calculate the most probable value of the height of each bench mark, assuming A to be 87.631 m AOD.

(TP. Ans. 126.882 m, 32.000 m, 52.037 m)

4 From the following table of recorded values in a crossed quadrilateral, derive the correlative normal equations using a tabulation process.

Angles: 1 71° 34′ 55″   2 39° 01′ 35″
         3 31° 28′ 32″   4 37° 54′ 51″
         5 24° 21′ 41″   6 86° 14′ 45″
         7 46° 43′ 44″   8 22° 39′ 48″

(Do not solve the equations.)

1.17 Comparison of the indirect and direct methods

In the **indirect method** (observational equations), the solution involves the normal equations,

$$A^T WAx = A^T Wl \qquad \text{(equation 1.38a)}$$

with its solution,

$$x = (A^T WA)^{-1}(A^T Wl) \qquad \text{(equation 1.39a)}$$

and the variance–covariance matrix

$$\sigma_{xx} = \sigma_0^2 (A^T WA)^{-1} \qquad \text{(equation 1.41)}$$

Here, the number of normal equations equals the number of unknowns. It is important that all the observational equations contain only *independent* unknowns.

Once the equations are written and transposed into normal equations, and the latter are solved, then an added advantage is the use of the variance–convariance matrix, which gives an indication of the precision of all the derived values and their interaction. This is of particular significance with the use of the 'variation of coordinates' method (see p. 43).

In the **direct method** (conditional equations), the solution involves 'correlative normal equations' of the form

$$(AW^{-1}A^T)k = (Y) \qquad \text{(equation 1.62)}$$

with the solution of the correlatives

$$k = (AW^{-1}A^T)^{-1}(Y) \qquad \text{(equation 1.63)}$$

and the back substitution for residuals

$$r = (W^{-1}A^T)(AW^{-1}A^T)^{-1}(Y) \qquad \text{(equation 1.64)}$$

Here, the number of equations to be solved is the same as the number of conditions, all of which must be independent. The latter is an important disadvantage in that some difficulty may arise in determining the number of conditions needed. On the other hand, conditions provide a check, as the absolute value (Y) indicates the amount by which the condition is not fulfilled. The variance–covariance matrix is also obtainable but is not so straightforward.

The ultimate decision as to which method is used may be dependent upon the computing facilities, when the size of the matrices governs the size of the stores needed. It would therefore be logical to suggest that for most engineering surveying purposes, where the method of variation of coordinates is becoming of prime importance, the indirect method is desirable if the computing facilities are available. If only hand calculators are available, then the direct method is preferred, particularly for network problems.

Example 1.19 Given the recorded data below, find the most probable values of the angles A, B and C: (a) by observational equations and (b) by conditional equations.

$$\begin{aligned} A &= 40°\,13'\,28.7'' \quad \text{wt. 1} \\ B &= 34°\,46'\,15.4'' \quad \text{wt. 1} \\ A+B &= 74°\,59'\,43.0'' \quad \text{wt. 2} \\ A+B+C &= 132°\,31'\,07.2'' \quad \text{wt. 1} \\ B+C &= 92°\,17'\,42.2'' \quad \text{wt. 3} \end{aligned}$$

(TP)

(a) Let $A = 40\,13'\,28.7'' + a$
$B = 34\,46'\,15.4'' + b$
$C = 57\,31'\,26.8'' + c$

Applying the weights to the observational equations gives

$$\begin{aligned} a &= 0 \\ b &= 0 \\ 2a + 2b\phantom{{}+c} + 2.20 &= 0 \\ a + b + c + 3.70 &= 0 \\ 3b + 3c &= 0 \end{aligned}$$

The normal equations then become

$$\begin{aligned} 4a + 3b + c + 5.9 &= 0 \\ 3a + 7b + 4c + 5.9 &= 0 \\ a + 4b + 4c + 3.7 &= 0 \end{aligned}$$

Solving these equations gives $a = -1.45''$, $b = 0.24''$, $c = -0.80''$. The most probable values are then

$$\begin{aligned} A &= 40°\,13'\,28.7'' - 1.45'' = 40°\,13'\,17.2'' \\ B &= 34°\,46'\,15.4'' + 0.24'' = 34°\,46'\,15.6'' \\ C &= 57°\,31'\,26.8'' - 0.80'' = 57°\,31'\,26.0'' \end{aligned}$$

(b) Let the errors in A, B, $\overline{A+B}$, $\overline{A+B+C}$ and $\overline{B+C}$ be r_1, r_2, r_3, r_4 and r_5 respectively. The conditional equations then become

$$\begin{aligned} r_1 + r_2 - r_3 &= -1.1'' \\ r_4 - r_5 - r_1 &= 3.7'' \end{aligned}$$

The least square condition will be

$$w_1 r_1^2 + w_2 r_2^2 + w_3 r_3^2 + w_4 r_4^2 + w_5 r_5^2 = \min.$$

The results may be tabulated as follows.

	$a \backslash k_1$	$b \backslash k_2$	$1/w$	aa/w	ab/w	bb/w
r_1	1	-1	1.00	1.00	-1.00	1.00
r_2	1		1.00	1.00		
r_3	-1		0.50	0.50		
r_4		1	1.00			1.00
r_5		-1	0.33			0.33
				2.50	-1.00	2.33

Thus
$$2.50k_1 - 1.00k_2 = -1.1''$$
$$-1.00k_1 + 2.33k_2 = 3.7''$$

Solving the normal correlative equations gives $k_1 = 0.24$ and $k_2 = 1.69$. Back substitution of k_1 and k_2 gives

$$r_1 = -1.45'' \text{ as above } r_2 = 0.24'' \text{ as above}$$
$$r_3 = -0.12'' \quad r_4 = 1.69'' \quad r_5 = -0.56''$$

Then
$$A = 40°\,13'\,27.2'' \text{ as before}$$
$$B = 34°\,46'\,15.6'' \text{ as before}$$
$$B + C = 92°\,17'\,41.6''$$

and
$$C = 57°\,31'\,26.0'' \text{ as before}$$

1.18 Adjustment by variation of co-ordinates (or differential displacements)

1.18.1 Introduction

This method may be used for all systems which involve the computations of coordinates, i.e. triangulation, trilateration, traversing, intersection, resection or such engineering surveying applications as deformation of structures. It is essentially a computer-based technique employing matrix algebra and relies on assumed coordinates of stations which have been provisionally computed with lengths and bearings (or angles) with error ratios not exceeding 1/4000 for the lengths and 1 minute of arc, when no further iteration is required.

1.18.2 General principles

Given provisional data for a line (IJ) of length S_{ij} and bearing ϕ_{ij} (see Fig. 1.9):

$$I(E_i, N_i); J; (E_j, N_j)$$

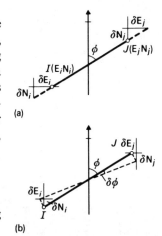

Fig. 1.9

Consider the length S_{ij}.

$$S_{ij}^2 = (E_j - E_i)^2 + (N_j - N_i)^2$$

Differentiating,

$$2S_{ij}dS_{ij} = 2(E_j - E_i)(dE_j - dE_i) + 2(N_j - N_i)(dN_j - dN_i)$$
$$dS_{ij} = (E_j - E_i)(dE_j - dE_i)/S + (N_j - N_i)(dN_j - dN_i)/S \quad (1.65)$$

If the errors are small then

$$dS \simeq \delta S = S^o - S^c = \text{observed} - \text{computed}$$

Then, as $\sin \phi = \Delta E/S$ and $\cos \phi = \Delta N/S$,

$$\delta S_{ij} = (\delta E_j - \delta E_i)\sin \phi_{ij} + (\delta N_j - \delta N_i)\cos \phi_{ij} \quad (1.66)$$

This may be written in general terms as

$$r_{ij} = -K_{ij}\delta E_i - L_{ij}\delta N_i + K_{ij}\delta E_j + L_{ij}\delta N_j - \delta S_{ij} \quad (1.67)$$

where

$$K_{ij} = \sin \phi_{ij} = \Delta E_{ij}/S_{ij} \quad \text{and}$$
$$L_{ij} = \cos \phi_{ij} = \Delta N_{ij}/S_{ij}$$

δS_{ij} = the difference between the observed and the computed length S_{ij}

r_{ij} = the residual error in the observational equation

$\delta E_i, \delta N_i, \delta E_j, \delta N_j$ = the adjustments to the provisional coordinates E_i, N_i, E_j, N_j.

Consider the bearing ϕ_{ij}.

$$\tan \phi_{ij} = (E_j - E_i)/(N_j - N_k)$$

Differentiating gives

$$\sec^2 \phi_{ij} d\phi_{ij} = [(N_j - N_i)(dE_j - dE_i)$$
$$- (E_j - E_i)(dN_j - dN_i)]/(N_j - N_i)^2$$
$$d\phi_{ij} = (N_j - N_i)(dE_j - dE_i)/S_{ij}^2$$
$$- (E_j - E_i)(dN_j - dN_i)/S_{ij}^2 \quad (1.68)$$

Then
$$\delta \phi_{ij} = (\Delta N_{ij}/S_{ij}^2)(\delta E_j - \delta E_i)$$
$$- (\Delta E_{ij}/S_{ij}^2)(\delta N_j - \delta N_i) \quad (1.69)$$

Generally the observational equations will be of the form

$$r_{ij} = -P_{ij}\delta E_i + Q_{ij}\delta N_i + P_{ij}\delta E_j - Q_{ij}\delta N_j - \delta \phi_{ij} \quad (1.70)$$

where $P_{ij} = 206\,265 \cos \phi_{ij}/S_{ij} = 206\,265\,\Delta N_{ij}/S_{ij}^2$
$Q_{ij} = 206\,265 \sin \phi_{ij}/S_{ij} = 206\,265\,\Delta E_{ij}/S_{ij}^2$
$\delta \phi_{ij} = (\phi^o - \phi^c)_{ij} = \text{observed} - \text{computed}$

Consider the angle $\theta_{i(jk)}$, i.e. station set at i, from station j to station k. If

$$r_{ij} = -P_{ij}\delta E_i + Q_{ij}\delta N_i + P_{ij}\delta E_j - Q_{ij}\delta N_j - (\phi^o - \phi^c)_{ij}$$

and $\quad r_{ik} = -P_{ik}\delta E_i + Q_{ik}\delta N_i + P_{ik}\delta E_k - Q_{ik}\delta N_k - (\phi^o - \phi^c)_{ik}$

then for the angle $\theta_{i(jk)}$

$$\begin{aligned}r_{i(jk)} = \delta E_i(P_{ij} - P_{ik}) + \delta N_i(Q_{ik} - Q_{ij}) - P_{ij}\delta E_j + Q_{ij}\delta N_j \\ + P_{ik}\delta E_k - Q_{ik}\delta N_k - (\theta^o - \theta^c)_{i(jk)}\end{aligned}$$

(1.71)

The general process is then as follows.

1) Assume or compute the approximate coordinates of all the stations.
2) Derive precisely ϕ^c, S^c and θ^c where ϕ^o, S^o and θ^o have been observed.
3) Set up an equation for each such observation.
4) Form and solve the normal equations for δE and δN at each station.
5) Compute the adjusted coordinates.
6) Test their reliability by computing the standard errors and the error ellipse data using the variance–covariance matrix (see pp. 21, 22).

Note the following points.

a) the minimum conditions to be fulfilled are
   (i) one fixed point is required to determine position;
   (ii) one bearing is fixed to determine orientation;
   (iii) one length is fixed to determine scale.
b) If a length S_{ij} is fixed then $\delta S_{ij} = 0$.
c) If a bearing ϕ_{ij} is fixed then $\delta\phi_{ij} = 0$.
d) If a point i is fixed, then $\delta E_i = \delta N_i = 0$.

The method is illustrated in Chapter 2.

Bibliography

ALLAN, A. L., HOLLWEY, J. R., and MAYNES, H. H. B., *Practical Field Surveying and Computations.* Heinemann (1968)

ASHKENAZI, V., The solution and error analysis of large geodetic networks. *Survey Reviews*, XIX, 146, 147 (1968)

ASHKENAZI, V., Adjustment of control networks for precise engineering networks. *Chartered Surveyor* (Jan. 1970)

ASHKENAZI, V., COUTIE, M. G., and SNELL, C., *Matrix Methods in Civil Engineering Networks.* Prepared for a Nottingham Seminar (1968)

ASHKENAZI, V., WUDDAH-MARTEY, E. E. L., and DODSON, A. H., Rigorous adjustment of an EDM traverse on a desk calculator. *Survey Review*, XXI, 165 (1972)

BIRD, R. G., Least square adjustment of EDM traverses. *Survey Review*, XXI, 165 (1972)

BRIGGS, N., *The Effects of Errors in Surveying*, Griffin

BRITISH STANDARD 2846 (1957)

COOPER, M. A. R., *Fundamentals of Survey Measurement and Analysis.* Crosby Lockwood Staples (1974)

JAMESON, A. H., *Advanced Surveying.* Pitman (1961)

LILLEY, J. E., Least square adjustment of dissimilar quantities. *Empire Survey Review*, XVI, 121 (1961)

MURCHISON, D. E., *Surveying and Photogrammetry.* Newnes-Butterworth (1977)

RAINSFORD, H. F., *Survey Adjustments and Least Squares.* Constable (1957)

RICHARDUS, P., *Project Surveying.* North-Holland (1965)

SCHOFIELD, M. W., Traverse adjustments by variation of coordinates. Papers in *The Civil Engineering Surveyor*, IV (1979)

THOMPSON, E. H., *An Introduction to the Algebra of Matrices with Some Applications.* Adam Hilger (1969)

2
Orientation

2.1 Methods of orientation

Surveys may be orientated using the following alternative meridians:

a) True North;   b) Grid North;
c) Magnetic North; d) Assumed North.

2.1.1 True North

True North may be determined by

a) astro-surveying (not dealt with here) *or*
b) gyroscopic observations (see p. 97).

2.1.2 Grid North

Grid North is used as the orientation for the OS coordinate system and, as many large construction projects are based upon grid coordinates, it is necessary to connect local surveys to the national survey. In the case of the coal mining industry there is a legal requirement to connect their surveys to the grid. The processes of connection involve intersection and resection techniques.

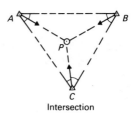
Intersection

2.1.3 Magnetic North

Magnetic North is seldom used for engineering surveying and is not considered here.

2.1.4 Assumed North

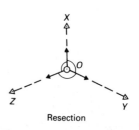
Resection

Assumed North is used extensively where relative values only are required.

2.2 Intersection and resection

These are processes for the fixing of subsidiary stations relative to previously determined coordinated points (see Fig. 2.1). **Intersection** relates to observations made from the fixed stations to the unfixed stations whilst **resection** is the reciprocal,

Fig. 2.1

47

i.e. observations are made from the unfixed stations to the fixed stations. In the case of intersection any two rays will produce a unique solution and where more than two rays are observed the most probable value must be derived, e.g. using least squares. For resection, any three rays will produce a unique solution.

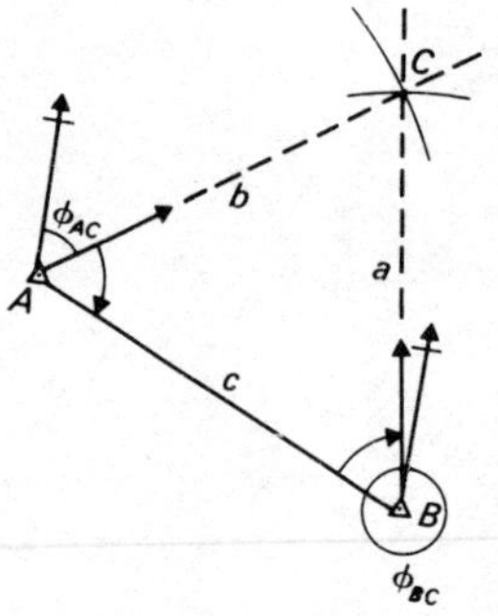

2.3 Intersection

The following methods will be discussed.

a) Two rays to fix a point.
b) Multi-ray processes involving (i) semi-graphic methods; (ii) computation by variation of coordinates.

Fig. 2.2

2.3.1 Two ray intersection methods (Fig. 2.2)

(a) *Intersection by bearings*
Given $A(E_A, N_A)$, $B(E_B, N_B)$ and bearings ϕ_{AC} and ϕ_{BC}, then

$$E_C = \frac{\Delta N_{AB} + E_A \cot \phi_{AC} - E_B \cot \phi_{BC}}{\cot \phi_{AC} - \cot \phi_{BC}} \qquad (2.1)$$

and $\quad N_C = \dfrac{\Delta E_{AB} + N_A \tan \phi_{AC} - N_B \tan \phi_{BC}}{\tan \phi_{AC} - \tan \phi_{BC}} \qquad (2.2)$

With the use of hand held calculators, it is generally preferable to use the partial coordinate formulae,

$$\Delta N_{AC} = (\Delta E_{AB} - \Delta N_{AB} \tan \phi_{BC})/(\tan \phi_{AC} - \tan \phi_{BC}) \quad (2.3)$$
and $\quad \Delta E_{AC} = \Delta N_{AC} \tan \phi_{AC} \qquad (2.4)$

(b) *Intersection by angles*
Given angles $A = CAB$, $B = ABC$, then

$$E_C = \frac{E_A \cot B + E_B \cot A - \Delta N_{AB}}{\cot A + \cot B} \qquad (2.5)$$

and $\quad N_C = \dfrac{N_A \cot B + E_B \cot A + \Delta E_{AB}}{\cot A + \cot B} \qquad (2.6)$

or $\quad \Delta E_{AC} = (\Delta E_{AB} \cot A - \Delta N_{AB})/(\cot A + \cot B) \qquad (2.7)$

and $\quad \Delta N_{AC} = (\Delta N_{AB} \cot A + \Delta E_{AB})/(\cot A + \cot B) \qquad (2.8)$

(c) *Intersection by lengths*
Triangular solution: Angles may be computed by (i) half angle formulae; (ii) the cosine rule.
(i) The half angle formulae method uses the formula

$$C = 2 \tan^{-1}\left[(s-a)(s-b)/s(s-c)\right] \quad \text{etc.} \qquad (2.9)$$

This is preferable to the other half angle formulae because of the arithmetical check

$$s = (s-a) + (s-b) + (s-c) = (a+b+c)/2$$

(ii) The cosine rule method uses the formula

$$C = \cos^{-1}(a^2 + b^2 - c^2)/2ab \quad \text{etc.} \quad (2.10)$$

In each case a final check is obtained with the summation of the angles, i.e.

$$A + B + C = 180 + E$$

where E is the spherical excess, only applicable when the lengths are greater than 20 km.

The coordinates of C are then computed as follows.

$$E_C = b \sin(\phi_{AB} - A) + E_A = a \sin(\phi_{BA} + B) + E_B$$
$$N_C = b \cos(\phi_{AB} - A) + N_A = a \cos(\phi_{BA} + B) + N_B$$

Direct solution of coordinates: This may be performed by the use of the intersection formulae (by angles), equation 2.5:

$$E_C = (E_A \cot B + E_B \cot A - \Delta N_{AB})/(\cot A + \cot B)$$

and

$$N_C = (N_A \cot B + N_B \cot A + \Delta E_{AB})/(\cot A + \cot B)$$
$$\text{(equation 2.6)}$$

As $\cos A = (b^2 + c^2 - a^2)/2bc$ (equation 2.10) and the area of the triangle

$$\Delta' = [s(s-a)(s-b)(s-c)]^{1/2} = \tfrac{1}{2} bc \sin A$$

then

$$\sin A = 2\Delta'/bc, \quad \cot A = (b^2 + c^2 - a^2)/4\Delta'$$

and $\quad \cot B = (a^2 + c^2 - b^2)/4\Delta'$

Then, substituting into equation 2.5,

$$E_C = \frac{E_A(a^2 + c^2 - b^2)/4\Delta' + E_B(b^2 + c^2 - a^2)/4\Delta' - \Delta N_{AB}}{[(b^2 + c^2 - a^2) + (a^2 + c^2 - b^2)]/4\Delta'}$$

$$= \frac{E_A(a^2 + c^2 - b^2) + E_B(b^2 + c^2 - a^2) - 4\Delta' \Delta N_{AB}}{2c^2}$$

$$= (E_A + E_B)/2 - \Delta E_{AB}(a^2 - b^2)/2c^2 - 2\Delta' \Delta N_{AB}/c^2 \quad (2.11)$$

Similarly,

$$N_C = (N_A + N_B)/2 - \Delta N_{AB}(a^2 - b^2)/2c^2 + 2\Delta' \Delta E_{AB}/c^2 \quad (2.12)$$

To aid the computation let

$$p = (a^2 - b^2)/2c^2$$

and $\quad q = 2\sqrt{[s(s-a)(s-b)(s-c)]}/c^2$

Then $\quad E_C = (E_A + E_B)/2 - p\Delta E_{AB} - q\Delta N_{AB} \quad (2.11\text{a})$

and $\quad N_C = (N_A + N_B)/2 - p\Delta N_{AB} + q\Delta E_{AB} \quad (2.12\text{a})$

Example 2.1 Given

	E (m)	N (m)
A	10 000.00	10 000.00
B	13 462.56	11 373.62

and $AC = 3051.63$ m $BC = 1980.27$ m

$\phi_{AB} = \tan^{-1}(3462.56/1373.62) = 68.3615°$

$S_{AB} = 1373.62 \sec 68.3615 = 3725.07$ m

Then

$$a = 1980.27 \quad s-a = 2398.215$$
$$b = 3051.63 \quad s-b = 1326.855$$
$$c = \underline{3725.07} \quad s-c = \underline{653.415}$$

and $2s = 8756.97$

$s = 4378.485 \qquad s = 4378.485$ (Check)

Then (by equation 2.9),

$$A = 2\tan^{-1}[(s-b)(s-c)/s(s-a)]^{1/2} = 32.063\,24$$
$$B = 2\tan^{-1}[(s-a)(s-c)/s(s-b)]^{1/2} = 54.890\,74$$
$$C = 2\tan^{-1}[(s-a)(s-b)/s(s-c)]^{1/2} = \underline{93.046\,02}$$
$$180.000\,00$$

or (by equation 2.10),

$$A = \cos^{-1}(b^2+c^2-a^2)/2bc = 32.063\,24$$
$$B = \cos^{-1}(a^2+c^2-b^2)/2ac = 54.890\,74$$
$$C = \cos^{-1}(a^2+b^2-c^2)/2ab = 93.046\,02 \quad \text{(as above)}$$

The coordinates of C are then

$E_C = 3051.63 \sin(68.3615 - 32.0632) + 10\,000.00$
$ = 11\,806.53$ m

$N_C = 3051.63 \cos(68.3615 - 32.0632) + 10\,000.00$
$ = 12\,459.45$ m

The same values are obtained using the line BC.
By equation 2.11a,

$$E_C = (E_A + E_B)/2 - p\,\Delta E_{AB} - q\,\Delta N_{AB}$$

By equation 2.12a,

$$N_C = (N_A + N_B)/2 - p\,\Delta N_{AB} + q\,\Delta E_{AB}$$

where $p = (a^2 - b^2)/2c^2$ and $q = 2\Delta/c^2$

Then $\quad a = 1980.27 \quad s-a = 2398.215 \quad b^2 = 9312\,445.66$
$\quad\quad\quad b = 3051.63 \quad s-b = 1326.855 \quad a^2 = 3921\,469.27$
$\quad\quad\quad\quad\quad\quad\quad\quad\quad\quad\quad\quad\quad\quad a^2 - b^2 = -5390\,976.39$
$\quad\quad\quad c = 3725.07 \quad s-c = 653.415 \quad c^2 = 13\,876\,146.50$
$\quad\quad\quad 2s = 8756.97$
$\quad\quad\quad s = 4378.485 \quad\quad\quad 4378.485$
$\quad\quad\quad \Delta' = [s(s-a)(s-b)(s-c)]^{1/2}$
$\quad\quad\quad\quad = 3017\,258.33$
$\quad\quad\quad q = 2\Delta/c^2 = 0.434\,884$
$\quad\quad\quad p = (a^2 - b^2)/2c^2 = -0.194\,253$

	E		N
A	10 000.00		10 000.00
B	13 462.56		11 373.62
ΔE_{AB}	3 462.56	ΔN_{AB}	1 373.62
ΣE	23 462.56	ΣN	21 373.62
$\Sigma E/2$	11 731.28	$\Sigma N/2$	10 686.81

Then $\quad E_C = 11\,731.28 - (-0.194\,253 \times 3462.56)$
$\quad\quad\quad\quad\quad - (0.434\,884 \times 1373.62)$
$\quad\quad\quad\quad = 11\,806.52\,\text{m}$
$\quad\quad N_C = 10\,686.81 - (-0.194\,253 \times 1373.62)$
$\quad\quad\quad\quad\quad + (0.434\,884 \times 3462.56)$
$\quad\quad\quad\quad = 12\,459.45\,\text{m}$

2.4 Multi-ray intersection

Where more than two rays intersect there is no unique solution and the most probable value must be derived by either a semi-graphic method or a least squares solution.

2.4.1 Semi-graphic intersection

The process is dependent upon a first approximation of the point to be fixed known as the 'trial point' P'. This point may be obtained either by estimation by plotting or by computation from two rays.

Rays are unlikely to pass through P' because of errors in:

a) the trial point coordinates;
b) the triangulation coordinates;
c) the angular observations.

The trial point coordinates are used as the axes of a graph and cutting points on the axes are computed using the equations for the 'cut' coordinates,

$$N_{P_2} = N_S + dN = N_S + \Delta E \cot \phi \qquad (2.13)$$

or $\quad E_{P_1} = E_S + dE = E_S + \Delta N \tan \phi \qquad (2.14)$

(see Fig. 2.3) where

E_P, N_P = the coordinates of the rays cut on the axes
E_S, N_S = the station coordinates
dE, dN = the partial coordinates to the cutting points
ϕ = the bearing of the ray
$\delta E, \delta N$ = the error in the estimate of P' based upon the ray.
$\Delta E, \Delta N$ = the partial coordinates $(E_{p'} - E_s)$, $(N_{p'} - N_s)$

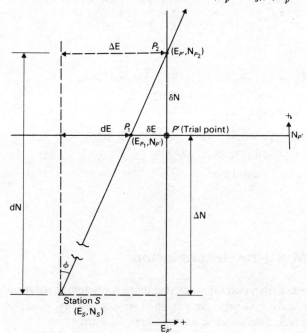

Fig. 2.3

The rays may now be plotted on the graph using

a) both cut coordinates; *or*
b) preferably one of the cut coordinates with the direction of the ray plotted by protractor. By this method it is desirable to choose either ΔE or ΔN (whichever is the larger), and

2.4 MULTI-RAY INTERSECTION 53

then a computed value dN or dE is derived, based upon the above equations. Based upon the larger value the trigonometrical value, i.e. $\cot \phi$ or $\tan \phi$, will be less than 1.

The method is now illustrated by an example.

Example 2.2 Observations from five triangulation stations A, B, C, D and E have been taken to fix the coordinates of station P. From the recorded data below, find the most probable value of the coordinates of P, by the semi-graphical method.

	E (m)	N (m)	Bearing to P
A	40 747.87	48 539.17	34° 01′ 01.4″
B	41 865.65	54 304.23	123° 15′ 03.1″
C	46 777.18	54 641.86	241° 51′ 58.2″
D	50 906.16	51 810.40	279° 49′ 46.2″
E	43 804.81	49 349.76	359° 44′ 29.9″

The provisional coordinates of P, as computed from rays A and B, are

    43 788.18 E   53 043.72 N   (RICS/M)

The coordinates of the 'cut' on the axes may be tabulated as follows.

Ray from station	C		D		E	
Bearing from station	241° 51′ 58.2″		279° 49′ 46.2″		359° 44′ 29.9″	
	E	N	E	N	E	N
Provisional coords	43 788.18	53 043.72	43 788.18	53 043.72	43 788.18	53 043.72
Station coords	46 777.18	54 641.86	50 906.16	51 810.40	43 804.81	49 349.76
(select larger) ΔE/ΔN	− 2989.00		− 7117.98			3693.96
$\cot\phi/\tan\phi$	0.534 709		− 0.173 260			− 0.004 509
dN/dE		− 1598.25		1233.26	− 16.66	
'Cut' coords		53 043.61		53 043.66	43 788.15	
Error		− 0.11		− 0.06	− 0.03	

For line CP,

1) $\Delta E > \Delta N$, therefore $dN = \Delta E \cot \phi$.
2) Cut coordinates $N_P = N_C + dN$.
3) Error = cut coords − provisional coords.

The graph is now plotted and represents the rays shown on Fig. 2.4 and its enlargement on Fig. 2.5.

It should be noted that the trial point coordinates are computed from rays A and B and thus, when plotted on the graph, pass through the axes of P'

The most probable value of the coordinates are now derived, based upon an **equal shift** method. Assuming all the rays on the graph have bearings subjected to equal error σ, then the amount of 'shift' required for each ray is proportional to its length (see

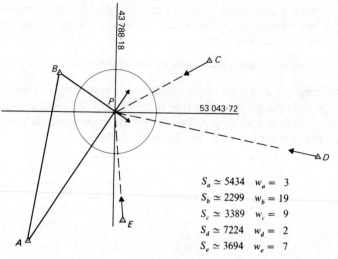

Fig. 2.4

$S_a \simeq 5434 \quad w_a = 3$
$S_b \simeq 2299 \quad w_b = 19$
$S_c \simeq 3389 \quad w_c = 9$
$S_d \simeq 7224 \quad w_d = 2$
$S_e \simeq 3694 \quad w_e = 7$

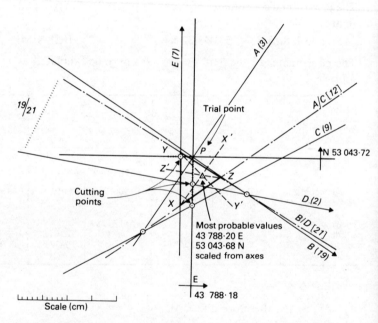

Fig. 2.5 Enlargement of Fig. 2.4

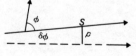

Fig. 2.6

Fig. 2.6), i.e.

$$\delta\phi = p/S \qquad (2.15)$$

where p is the perpendicular offset and s is the length of the ray. As $\delta\phi$ is small then

$$\delta\phi'' = 206\,265\, p/S \qquad (2.15a)$$

and the adjusted ray on the graph is assumed parallel to the original plotted ray. Then $p \propto S$.

The following procedure may be used to find the most probable value on the graph.

1) Give each ray a weight (w). As $\sigma \propto p \propto S$; $w \propto 1/\sigma^2 \propto 1/S^2$.
2) Select the two rays which intersect most acutely.
3) Subdivide the acute angle, inversely proportional to the weights, to provide a resultant mean ray.
4) Continue the process until only three rays remain, forming a final triangle of error.
5) Subdivide the apexes of the triangle simultaneously, thus providing a trisection, the most probable value.
6) Scale off the most probable value of the coordinates of P.
7) The most probable value of the bearings of the rays may be derived by applying the adjustment $\delta\phi$ based upon equation 2.15a.

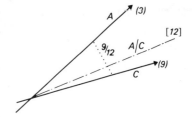

Fig. 2.7

With reference to the example the following points should be noted (see Fig. 2.7).

1) Ray CP cuts the axis at -11 cm N, and the bearing of the ray, say 241.9°, passes through this point.
2) The weight attributed to the ray $\propto 1/S^2$, e.g. $w_A \propto 1/5430^2$, say 3×10^{-8}. The proportional values are then $A(3)$, $B(19)$, $C(9)$, $D(2)$ and $E(7)$.
3) Rays A and C are chosen as the most acute and are thus subdivided in the ratio 9:3 producing a resultant ray A/C, weight 12. Rays B and D are similarly subdivided to give B/D (21).
4) The final triangle of error XYZ remains, and the subdivision of the apexes gives lines XX', YY' and ZZ' which trisect at P, the most probable value.
5) The errors in the estimate (δE, δN) are read off from the graph to give:

$E_P = 43\,788.180 + 0.024 = 43\,788.204$ say $43\,788.20$ m E

$N_P = 53\,043.720 - 0.043 = 53\,043.677$ say $53\,043.68$ m N

2.4.2 Intersection by variation of coordinates

By equation 1.70,

$$\delta\phi_{ij} + r_{ij} = -P_{ij}\delta E_i + Q_{ij}\delta N_i + P_{ij}\delta E_j - Q_{ij}\delta N_j$$
$$= (\phi^o - \phi^c)_{ij} + r_{ij}$$

If station (i) is fixed, i.e. the triangulation station, then $\delta E_i = \delta N_i = 0$. The equation then becomes

$$r_{ij} = P_{ij}\delta E_j - Q_{ij}\delta N_j - (\phi^o - \phi^c)_{ij}$$

which when transposed into the general observational equation is

$$r_i = a_i \delta E_p + b_i \delta N_p - l_i \qquad (2.16)$$

where $\quad a_i = 206\,265 \cos\phi_{ij}/S_{ij} \quad = 206\,265\,\Delta N_{ij}/S_{ij}^2$
$\qquad b_i = -206\,265 \sin\phi_{ij}/S_{ij} = -206\,265\,\Delta E_{ij}/S_{ij}^2$
$\qquad l_i = \phi_{ij}^c - \phi_{ij}^o$

Example 2.3 Compute the previous example by the method of 'variation of coordinates'.

Using equation 2.16,

$$r_i = a_i \delta E_p + b_i \delta N_p - l_i$$

where $\quad a_i = P_{ij} = 206\,265 \cos\phi_{ij}/S_{ij}$
$\qquad b_i = -Q_{ij} = -206\,265 \sin\phi_{ij}/S_{ij}$

Let $i \equiv A, B, C, D$ and $E; j \equiv P$. Note that all the stations are fixed except P.

	E_{ij}	N_{ij}	ϕ^o	ϕ^c	$l = \delta\phi''$	S_{ij}	$a = P_{ij}$	$b = -Q_{ij}$
AP	3040.31	4504.55	34° 01' 01.4	01.44"	−0.04	5434.56	31.459	−21.233
BP	1922.53	−1260.51	123° 15' 03.1	03.31"	−0.21	2298.91	−49.196	−75.033
CP	−2989.00	−1598.14	241° 51' 58.2	2 /03.91"	−5.71	3389.42	−28.695	53.665
DP	−7117.98	1233.32	279° 49' 46.2	47.80	−1.60	7224.04	4.874	28.133
EP	− 16.63	3693.96	359° 44' 29.9	31.41	−1.51	3694.00	55.837	0.252

Then $\quad [aa] = 7374.845 \quad [ab] = 1634.628$
$\qquad -[al] = -80.809$
$\qquad [bb] = 9752.253 \quad -[bl] = 335.215$
$\qquad [ll] = 37.490$

The normal equations are then

$$7374.845\,\delta E + 1634.628\,\delta N - 80.809 = 0$$
$$1634.628\,\delta E + 9752.253\,\delta N + 335.215 = 0$$

Solving these equations using equations 1.48 and 1.49,

$$\delta E = 0.019\,293 \text{ m} \quad \text{and} \quad \delta N = -0.037\,607 \text{ m}$$

The most probable values for P are then

$\qquad 43\,788.180 + 0.019 = $ say $43\,788.20$ m E
$\qquad 53\,043.720 - 0.038 = 53\,043.68$ m N

Applying the values of δE and δN to equation 2.16,

$$r_i = a_i \delta E_p + b_i \delta N_p - l_i$$

then $\quad r_a = (31.459 \times 0.019) + (-21.233 \times -0.038) + 0.04$
$\qquad\quad = 1.44$

Similarly,

$\qquad r_b = 2.13 \quad r_c = 3.13 \quad r_d = 0.62 \quad r_e = 2.56 \quad$ and
$[rr] = 23.35$

2.4 MULTI-RAY INTERSECTION

By equation 1.42,

$$\sigma_0^2 = [rr]/(m-n) = 23.35/3 = 7.78$$

The same result may be obtained by using the summation values. As

$$r_i = a_i\delta E + b_i\delta N - l_i$$
$$r_i^2 = a_i^2\delta E^2 + 2a_ib_i\delta E\delta N - 2a_il_i\delta E - 2b_il_i\delta N + b_i^2\delta N^2 + l_i^2$$

Therefore

$$[rr] = [aa]\delta E^2 + 2[ab]\delta E\delta N - 2[al]\delta E - 2[bl]\delta N$$
$$+ [bb]\delta N^2 + [ll]$$
$$= \delta E([aa]\delta E + [ab]\delta N + [-al])$$
$$+ \delta N([ab]\delta E + [bb]\delta N + [-bl])$$
$$+ [-al]\delta E + [-bl]\delta N + [ll]$$

Then

$$[rr] = -[al]\delta E - [bl]\delta N + [ll] \qquad (2.17)$$

as $[aa]\delta E + [ab]\delta N - [al] = [ab]\delta E$
$+ [bb]\delta N - [bl] = 0$

Therefore

$$\sigma_0^2 = \{-[al]\delta E - [bl]\delta N + [ll]\}/(m-n) \qquad (2.18)$$
$$= (-80.809 \times 0.019\,293) + (335.215 \times -0.037\,607)$$
$$+ 37.490/3$$
$$= 23.325/3 = 7.78 \text{ as above.}$$

Thus by equation 1.51,

$$\sigma_E = \sigma_0\{[bb]/K\}^{1/2} = \pm 0.033 \text{ m}$$

by equation 1.52,

$$\sigma_N = \sigma_0\{[aa]/K\}^{1/2} = \pm 0.029 \text{ m}$$

The values of P are thus $43\,788.20 \pm 0.03$ m E, $53\,043.68 \pm 0.03$ m N.

The above calculations show the tedium using a hand held calculator. By contrast the computer using matrix algebra tackles the problem as follows.

By the basic matrix equation $r = Ax - l$,

$$r = \begin{pmatrix} 31.459 & -21.233 \\ -49.196 & -75.033 \\ -28.695 & 53.665 \\ 4.874 & 28.133 \\ 55.837 & 0.252 \end{pmatrix} \begin{pmatrix} \delta E \\ \delta N \end{pmatrix} - \begin{pmatrix} -0.04 \\ -0.21 \\ -5.71 \\ -1.60 \\ -1.51 \end{pmatrix}$$

To solve these equations,

$$x = (A^T A)^{-1}(A^T l)$$

where $(A^TA) = \begin{pmatrix} 7374.8446 & 1634.6285 \\ 1634.6285 & 9752.2528 \end{pmatrix}$

$$(A^TA)^{-1} = \begin{pmatrix} 1.4083 & -0.2360 \\ -0.2360 & 1.0650 \end{pmatrix} \times 10^{-4}$$

$$(A^Tl) = \begin{pmatrix} 80.809 \\ -335.215 \end{pmatrix}$$

Note that the normal equations are given as $(A^TAx) - (A^Tl)$. Compare with the other solution.

Then $x = \begin{pmatrix} \delta E \\ \delta N \end{pmatrix} = \begin{pmatrix} 0.0193 \\ -0.0376 \end{pmatrix}$

Using $r = Ax - l$, the residuals are given by

$$r = \begin{pmatrix} 1.4454 \\ 2.0826 \\ 3.1382 \\ 0.6360 \\ 2.5778 \end{pmatrix}$$

$$[rr] = r^Tr = 23.3243$$
$$\sigma_0^2 = r^Tr/(m-n)$$
$$= 23.3243/3 = 7.775$$

The variance–covariance matrix is then given by

$$\sigma_{xx} = \sigma_0^2 (A^TA)^{-1}$$
$$= \begin{pmatrix} \sigma_E^2 & \sigma_{EN} \\ \sigma_{EN} & \sigma_N^2 \end{pmatrix} = \begin{pmatrix} 0.00109 & -0.00018 \\ -0.00018 & 0.00083 \end{pmatrix}$$

so $\sigma_E = \sqrt{0.00109} = \pm 0.0331$ m

$\sigma_N = \sqrt{0.00083} = \pm 0.0288$ m

$\sigma_{EN} = -0.00018$

By equation 1.46, the error ellipse values are $\sigma_{max} = 0.0344$, $\sigma_{min} = 0.0270$, $\phi_m = 116.9°$.

2.4.3 Semi-graphic intersection using linear values of the rays

As in other methods, the provisional coordinates of the 'trial point' P' are derived by either (a) estimation by plotting, or (b) computation using two rays.

By comparison with the angular methods, the error is assumed to be due to the linear value and thus the position of P must lie on an arc of radius equal to the measured length (Fig. 2.8(a)). At a small scale, used for plotting the graph, the arc may be assumed to be a straight line, lying at right angles (i.e. a

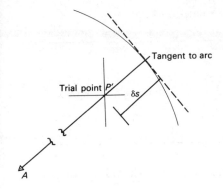

Fig. 2.8(a)

tangent) to the ray (which passes through the trial point P'), and at a distance from it equal to the difference between the measured and the computed length, i.e. $S^o - S^c = \delta S$. An error figure is thus produced and the most probable position must lie at the 'centre of gravity' of all points produced by intersecting rays. The method is illustrated using the previous example assuming the lengths are measured as follows.

$AP = 5434.561$, $BP = 2298.914$, $CP = 3389.400$
$DP = 7224.016$, $EP = 3694.015$

Assume the provisional coordinates of P are 43 788.18 m E, 53 043.72 m N, derived from the rays AP and BP (see Fig 2.8 (b)).

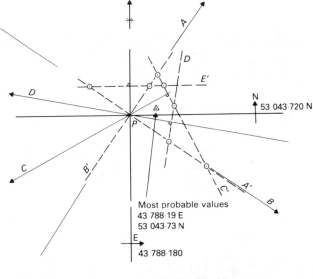

Fig. 2.8(b)

The results may be tabulated as follows.

	ϕ^c	S^o	S^c	δS
AP	34.0	5 434.561	5 434.561	0.000
BP	123.3	2 298.914	2 298.914	0.000
CP	241.9	3 389.400	3 389.421	−0.021
DP	279.8	7 224.016	7 224.037	−0.021
EP	359.7	3 694.015	3 693.997	0.018

In Fig. 2.8(b), the following points should be noted.
1) Rays are plotted through the axis P.
2) Tangents plotted --- at right angles to rays, δs from axis.
3) Most probable value derived by inspection.

2.4.4 Variation of coordinates method using linear values

Applying the variation of coordinates method to the last example (see p. 59), since one end of the line is fixed, the general equation is given as

$$r_i = a_i \delta E_j + b_i \delta N_j - l_i$$

where $a_i = K_{ij} = \sin\phi_{ij} = \Delta E_{ij}/S_{ij}$
$b_i = L_{ij} = \cos\phi_{ij} = \Delta N_{ij}/S_{ij}$

The results may be tabulated as follows.

	ΔE_{ij}	ΔN_{ij}	$a_j = K_{ij}$	$b_j = L_{ij}$	S^o	S^c	$l_j = \delta S_{ij}$
AP	3040.31	4504.55	0.9594	0.8289	5434.561	5434.561	0.000
BP	1922.53	−1260.51	0.8363	−0.5483	2298.914	2298.914	0.000
CP	−2989.00	−1598.14	−0.8819	−0.4715	3389.400	3389.421	−0.021
DP	−7117.98	1233.32	−0.9853	0.1707	7224.016	7224.037	−0.021
EP	− 16.63	3693.96	−0.0045	1.0000	3694.018	3693.997	0.018

Then

$$(A^T A) = \begin{pmatrix} 3.3683 & 0.5798 \\ 0.5798 & 2.2392 \end{pmatrix}$$

$$(A^T A)^{-1} = \begin{pmatrix} 0.31072 & -0.08046 \\ -0.08046 & 0.46743 \end{pmatrix} \quad (A^T l) = \begin{pmatrix} 0.0400 \\ 0.0248 \end{pmatrix}$$

$$x = \begin{pmatrix} \delta E \\ \delta N \end{pmatrix} = \begin{pmatrix} 0.0104 \\ 0.0084 \end{pmatrix}$$

$$r = \begin{pmatrix} 0.0169 \\ 0.0041 \\ 0.0088 \\ 0.0121 \\ -0.0097 \end{pmatrix}$$

and $\sigma_0^2 = r^T r/(m-n) = 0.000\,21$

$\sigma_E = \pm 0.0079\,\text{m} \qquad \sigma_N = \pm 0.0097\,\text{m}$

$\sigma_{EN} = -0.000\,016$

$\sigma_{max} = 0.0100\,\text{m} \qquad \sigma_{min} = 0.0074 \qquad \phi_m = 157.5°$

The most probable values are then $43\,788.190 \pm 0.007\,\text{m}$ E, $53\,043.728 \pm 0.009\,\text{m}$ N.

Exercises 2.1

1 Observations from five triangulation stations A, B, C, D and E have been taken to fix the position of a station P. The coordinates of the stations and the azimuth of the intersecting rays are given as follows.

	E (m)	N (m)	Bearing to P	Distance to P (m)
A	300 747.87	298 539.17	034° 01′ 00″	5430
B	301 865.65	304 304.23	123° 15′ 03″	2300
C	306 777.18	304 641.86	241° 51′ 59″	3390
D	310 906.16	301 810.40	279° 49′ 47″	7220
E	303 804.81	299 349.76	359° 44′ 36″	3690

The provisional coordinates of P, as computed from rays AP and BP, are $303\,788.15\,\text{m}$ E, $303\,043.74\,\text{m}$ N.

Determine, by the semi-graphic method, the most probable values of the coordinates of the station P.

(Ans. $303\,788.26\,\text{m}$ E, $303\,043.67\,\text{m}$ N)

2 Given the following data, compute the most probable value of the coordinates of D, using the semi-graphic method.

	E (m)	N (m)	Lengths to D (m)
A	1000.00	3000.00	5198.43
B	2000.00	6000.00	4569.08
C	3000.00	1000.00	4309.12
D	6105.00	3990.00	(Assumed values)

(Ans. $6103.21\,\text{m}$ E, $3989.85\,\text{m}$ N)

3 Using the variation of coordinates method applied to intersection by bearings, for the data given below, compute the most probable value of the coordinates of C.

	E (m)	N (m)	Observed bearings
A	10 000.00	10 000.00	AC 010° 14′ 51″
B	13 359.87	9288.79	BC 322° 31′ 44″
D	11 681.31	11 098.68	DC 328° 34′ 20″
C′	10 537.16	12 971.27	(Computed from AC and BC)

(Ans. $10\,537.13\,\text{m}$ E, $12\,971.15\,\text{m}$ N)

4 Using the same control stations above, the following distances to C were measured. $AC = 3019.425$ m, $BC = 4639.874$ m, $DC = 2194.460$ m. Compute the most probable values of the coordinates of C.

(Ans. 10 537.14 m E, 12 971.26 m N)

5 EDM measurements, as shown below, have been made between the three triangulation stations A, B, C and P, a point on a structure which has previously been determined as having coordinates 439 776.000 m E, 392 847.000 m N.

	Grid coordinates		Grid length to P	
	E	N		
A	437 214.684	392 167.314	2649.960 m	± 7 mm
B	439 246.398	393 249.123	664.976 m	± 5 mm
C	440 116.746	389 114.886	3747.645 m	±10 mm

Using the variation of coordinates method, calculate the most probable values of the movement of the structure at this point assuming the first coordinates were correct.

(TP Ans. $\delta E + 2.5$ mm, $\delta N - 4.8$ mm)

2.5 Weighting of physically dissimilar quantities

When an adjustment comprises a combination of angular and linear values, it is essential to use dimensional weight factors (u). As $w \propto 1/s_{\bar{x}_2}$, then the adjustment of the observation to one of unit weight is obtained by multiplying by $\sqrt{w}$. Then $u = \sqrt{w} = 1/s_{\bar{x}}$.

If an observational equation (in seconds or metres) is multiplied by u the equation becomes dimensionless. For example, if $s_\phi = \pm 4''$ then $u_\phi = 1/4 = 0.25$, and if $s_S = \pm 0.04$ m then $u_s = 1/0.04 = 25$. Thus if s_ϕ is compatible with s_S, then the observational equations are respectively multiplied by 1 and 100.

It may be assumed that if the standard errors have been assigned correctly; then $\sigma_0^2 \simeq s_0^2$ (the unbiased variance) should equal unity. As given before, by equation 1.42,

$$\sigma_0^2 = [rr]/(m-n) \qquad (2.19)$$

where m is the number of observational equations and n is the number of unknowns.

Ashkenazi suggests that an answer K larger than unity implies that the standard errors have been underestimated by the factor K and if the process is then repeated with the new estimates the new σ_0^2 should approach unity. The following points should be noted.

2.5 WEIGHTING OF PHYSICALLY DISSIMILAR QUANTITIES

1) A large difference from unity suggests a gross error whilst a small difference $0.5 < \sigma_0 < 2.0$ indicates a poor estimate and a new value $s' = s \times \sigma_0$.
2) If all the observations consist of similar quantities, i.e. all angular or all linear, the unitless equations will produce the same solutions, whatever the weight factor, but a different σ_0^2. If a combination exists, then any change in the weight factors will change the solutions.
3) Unless there is a considerable change in the standard errors related to the same physical dimensions they may, for all practical purposes, be assumed the same, although dissimilar units will take very dissimilar weight factors.

The following procedure is suggested.

1) Using the angular observational equations an initial solution is obtained; i.e. δE, δN and s'_0 are computed.
2) Apply the weight factor $u' = 1/s'_0$ and compute the residuals $r_i = a_i \delta E + b_i \delta N - l_i$; the unbiased variance then becomes

$$\sigma_0^2 = [(r/s'^2_0)]/(m-n) \simeq \sigma_0^2/\sigma_0^2 = 1$$

3) Repeat the computations using u' for the final estimate of the standard error but now include the linear equations with the initial estimate of their standard errors $s_{S'}$. Solve for δE, δN and σ_0^2.
4) Any departure of σ_0^2 from unity is now assumed due to the estimate of the s_S.
5) Compute $s'_S = s_S \times \sigma_0$ and apply the new u to the linear observational equations for a new combined solution.
6) Reiterate this procedure until σ_0^2 is sufficiently close to unity.

Note that the value r/s may be used as a rejection criteria: if $r/s > 2.5$ for 100 observations or if $r/s > 2.0$ for less than 25 observations.

Example 2.4 Given the following data, compute the most probable value of the coordinates of D.

	E (m)	N (m)	Observed bearings to D	Observed distances to D
A	10 000.00	10 000.00	010° 14′ 51″	3019.425 m
B	13 359.87	9 288.79	322° 31′ 44″	4639.874 m
C	11 681.31	11 098.68	328° 34′ 20″	2194.460 m
D	10 537.16	12 971.27	(Assumed coordinates of D)	

Assuming the standard error in the bearings is $\pm 5''$ and the lines were measured with an EDM instrument which has a standard error of ±5 mm and a proportional error of 2 ppm compute the most probable coordinates of D.

This adjustment can only be adequately dealt with as part of a network adjustment carried out on a computer and following the principles laid down on pp. 62, 63.

The following is the type of output possible from a suitably written computer program.

Station coordinates

Station	Eastings	Northings	
1	10 000.0000	10 000.0000	(fixed)
2	13 359.8700	9288.7900	(fixed)
3	11 681.3100	11 098.6800	(fixed)
4	10 537.1600	12 971.2700	

Observations

	Observation no.	Stations concerned		Standard error	Observed value
Distances	1	1	4	0.011 m	3019.4250
	2	2	4	0.009 m	2194.4600
	3	3	4	0.014 m	4639.8740
Bearings	4	1	4	5.000 "	10° 14' 51.00"
	5	2	4	5.000 "	322° 31' 44.00"
	6	3	4	5.000 "	328° 34' 20.00"

Final adjusted coordinates

Station	Eastings	Northings
4	10 537.1391	12 971.2583

Standard errors in adjusted coordinates

0.0148 m   0.0079 m

Observation no.	Residual
1	−0.0054 m
2	0.0049 m
3	−0.0051 m
4	−1.2113 "
5	−0.7886 "
6	8.2046 "

Standard errors of adjusted distances and bearings

Station numbers	Standard error of distance	Standard error of bearing
1 to 2	Both stations are fixed	
1 to 3	Both stations are fixed	
1 to 4	0.0095 m	0.94"
2 to 3	Both stations are fixed	
2 to 4	0.0076 m	0.66"
3 to 4	0.0069 m	1.44"

Error ellipse

Station number	Max. standard error	Min. standard error	Bearing of major axis
4	0.0156 m	0.0063 m	249.9

This program had two reiterations with a final standard error $(\sigma_0) = 0.926$.

2.6 Resection

Four alternative methods are considered:

1) two point resection;
2) three point resection;
3) semi-graphic resection;
4) resection by variation of coordinates.

2.6.1 Two point resection

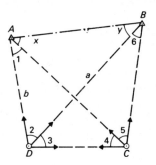

Fig. 2.9

The coordinates of the fixed points A and B are known (see Fig. 2.9). Angles 2, 3, 4 and 5 are measured. The coordinates of stations C and D are required.

Two methods of solution are considered here.

(a) By assuming a length CD and solving the triangles in appropriate sequence, the angles X and Y may be obtained. By intersection formulae or by solving triangles the coordinates of C and D may be computed.

(b) By derivation of a formula. In the quadrilateral $ABCD$,

$$X = (3+4) - Y$$

and $\quad \sin X = \dfrac{\sin 1 \sin 3 \sin 5 \sin Y}{\sin 2 \sin 4 \sin 6} \quad$ (by equation 2.32)

Combining these two equations,

$$\sin(3+4)\cos Y - \cos(3+4)\sin Y = \dfrac{\sin 1 \sin 3 \sin 5 \sin Y}{\sin 2 \sin 4 \sin 6}$$

Dividing throughout by $\sin(3+4)\sin Y$ and rearranging,

$$\cot Y = \dfrac{\sin 1 \sin 3 \sin 5}{\sin 2 \sin 4 \sin 6 \sin(3+4)} + \cot(3+4) \qquad (2.20)$$

This lends itself to either logarithmic or machine solution.

Example 2.5 Given the following data, it is required to compute the coordinates of C and D.

	E	N	Observed angles	
A	8 568.24	10 360.73	ADB (2)	57° 56′ 32″
			DCA (4)	32° 12′ 55″
B	12 423.80	12 606.60	BDC (3)	46° 51′ 19″
			ACB (5)	57° 56′ 32″

Derived angles (1) = (6) = 42° 59′ 14″. It can be seen that in view of the agreement between the above and the fact that (2) = (5) that the figure is a cyclic quadrilateral and thus (X) and (Y) agree with angles (3) and (4) respectively. Ignoring this fact it is possible to illustrate the two methods, as follows.

Method (a) Let $DC = 1$ (Fig. 2.9). Then
In $\triangle ADC$, $AD = b = DC \sin(4)/\sin(1) = 0.781\,863$
In $\triangle CDB$, $BD = a = DC \sin(4+5)/\sin(6) = 1.466\,624$
In $\triangle ABD$,

$$\tan(A-B)/2 = \frac{(a-b)}{(a+b)} \tan(A+B)/2$$

$$(A-B)/2 = \tan^{-1}(0.684\,761 \tan 61.0289°)/2.248\,487$$

$$= 28.8136°$$

$$(A+B)/2 = 61.0289°$$

By adding,

$$A = 89.8425° \quad \text{and} \quad X = (A) - (1)$$
$$= 46° 51′ 19″$$

By subtracting,

$$B = 32.2153° = Y = 32° 12′ 55″$$

From the given coordinates,

$$\Delta E_{AB} = 3855.56\,\text{m} \quad \text{and} \quad \Delta N_{AB} = 2245.87\,\text{m}$$

By equation 2.5, modified for relativity,

$$E_C = \frac{E_A \cot(Y+6) + E_B \cot X + \Delta N_{AB}}{\cot(Y+6) + \cot X}$$

$$= 13\,445.41\,\text{m}$$

By equation 2.6,

$$N_C = \frac{N_A \cot(Y+6) + N_B \cot X - \Delta E_{AB}}{\cot(Y+6) + \cot X}$$

$$= 8903.59\,\text{m}$$

Similarly $E_D = 9987.60\,\text{m}$ and $N_D = 7939.40$.

Method (b) By equation 2.20,

$$\cot Y = \frac{\sin(1) \sin(3) \sin(5)}{\sin(2) \sin(4) \sin(6) \sin(3+4)} + \cot(3+4)$$

If $P = [\sin(1) \sin(3) \sin(5)]/[\sin(2) \sin(4) \sin(6) \sin(3+4)]$ this may be derived using a good calculator. Alternatively,

$$\log P = \log \sin(1) + \log \sin(3) + \log \sin(5) - \log \sin(2)$$
$$- \log \sin(4) - \log \sin(6) - \log \sin(3+4)$$

Working in seconds, Shortredes tables of logarithmic trigonometrical functions provides the simpler solution. In both cases $Y = 32°\,12'\,55''$ and thus $X = 46°\,51'\,19''$ as before.

2.6.2 Three point resection

Given the coordinates of three fixed stations A, B and C, observations are taken at O giving angles, $AOB\,(\alpha)$ and $BOC\,(\beta)$.

(a) *Graphical solution by Collins' point method* (Fig. 2.10)
Case I The observer's position is outside triangle ABC. The procedure is as follows.

1) Plot the coordinates of A, B and C.
2) On line AC plot α at C and β at A to intersect at P (Collins' point).
3) Construct a circle to pass through A, P and C.
4) Join PB (or $B'P$) and produce to cut the circle at O (the position of the observer).

Note the following points.

1) $APCO$ is a cyclic quadrilateral.
   $AOP = ACP = AOB = \alpha$
   $POC = PAC = BOC = \beta$
2) In Case II where the observer is inside the $\triangle ABC$.
   $AOB = \alpha \quad PCA = 180 - \alpha = POA$
   $BOC = \beta \quad CAP = 180 - \beta = COP$
3) As P approaches B the solution becomes unstable and when $P = B$ there is no solution and B is said to lie on the 'danger circle' $ABCO$.

(b) *Mathematical solution by Blunt's method*
There are many solutions, but the one considered here is the one which uses the principles of graphical solution and which is also based upon the intersection formulae by angles and bearings. The method is based upon the principles of the Collins' point method.

1) In $\triangle APC$, using the intersection by angles, compute the coordinates of P.
2) In $\triangle ACO$, using the intersection by bearings, the coordinates of O may be computed if necessary.

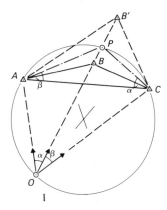
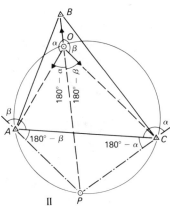

Fig. 2.10

In $\triangle APC$ (Fig. 2.10, Case I)

$$E_P = (E_A \cot \alpha + E_C \cot \beta - N_C + N_A)/(\cot \alpha + \cot \beta)$$
and $N_P = (N_A \cot \alpha + N_C \cot \beta + E_C - E_A)/(\cot \alpha + \cot \beta)$
$\phi_{PB} = \phi_{BO} = \tan^{-1}(E_B - E_P)/(N_B - N_P)$

$$= \tan^{-1} \frac{E_B(\cot \alpha + \cot \beta) - (E_A \cot \alpha + E_C \cot \beta - N_C + N_A)}{N_B(\cot \alpha + \cot \beta) - (N_A \cot \alpha + N_C \cot \beta + E_C - E_A)}$$

$$= \tan^{-1} \frac{\Delta E_{AB} \cot \alpha + \Delta E_{CB} \cot \beta + \Delta N_{AC}}{\Delta N_{AB} \cot \alpha + \Delta N_{CB} \cot \beta - \Delta E_{AC}} \qquad (2.21)$$

In Fig. 2.10,

$$\phi_{AO} = \phi_{BO} - \alpha \quad \text{and} \quad \phi_{CO} = \phi_{BO} + \beta$$

In $\triangle ACO$, by equation 2.3,

$$\Delta N_{AO} = \frac{\Delta E_{AC} - \Delta N_{AC} \tan \phi_{CO}}{\tan \phi_{AO} - \tan \phi_{CO}} \tag{2.22}$$

or

$$\Delta N_{CO} = \frac{\Delta E_{AC} - \Delta N_{AC} \tan \phi_{AO}}{\tan \phi_{CO} - \tan \phi_{AO}} \tag{2.22a}$$

and by equation 2.4,

$$\Delta E_{AO} = \Delta N_{AO} \tan \phi_{AO} \tag{2.23}$$

or

$$\Delta E_{CO} = \Delta N_{CO} \tan \phi_{CO} \tag{2.23a}$$

Then
$E_O = E_A + \Delta E_{AO}$ or $E_O = E_C + \Delta E_{CO}$
$N_O = N_A + \Delta N_{AO}$ or $N_O = N_C + \Delta N_{CO}$

As P approaches B an unstable solution exists and when P coincides with B they are said to lie with O on the 'danger circle'.

Applied to the observer's position inside the $\triangle ABC$ (Fig. 2.10, case II), this is obviously the better solution as no 'danger circle' is possible. In this case α and β are replaced throughout by $(180 - \alpha)$ and $(180 - \beta)$. This has the effect of changing all the trigonometrical functions and the resulting equation becomes

$$\phi_{BP} = \phi_{BO} = \tan^{-1} \frac{\Delta E_{AB} \cot \alpha + \Delta E_{CB} \cot \beta + \Delta N_{AC}}{\Delta N_{AB} \cot \alpha + \Delta N_{CB} \cot \beta - \Delta E_{AC}}$$

This is the same basic equation with the observer outside the $\triangle ABC$, except that the bearing is reversed, i.e. ϕ_{BP} appears instead of ϕ_{PB} but both equal ϕ_{BO}. This is not important as in the subsequent equations the trigonometrical functions are tangents and $\tan \phi = \tan (180 - \phi)$. The remainder of the solution follows the same pattern.

Example 2.6 An example of Blunt's resection. Given the following data, compute the coordinates of the observer's position (Fig. 2.11).

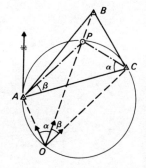

Fig. 2.11

	E (m)	N (m)	
A	904 365.74	428 373.70	$\alpha = 35° 41' 40''$
B	906 407.63	430 117.20	$\beta = 24° 59' 55''$
C	907 900.99	429 140.56	

By equation 2.21,

$$\phi_{PB} = \tan^{-1} \frac{\Delta E_{AB} \cot \alpha + \Delta E_{CB} \cot \beta + \Delta N_{AC}}{\Delta N_{AB} \cot \alpha + \Delta N_{CB} \cot \beta - \Delta E_{AC}}$$

From the above coordinates, the following data may be obtained.

	$\Delta E(m)$	$\Delta N(m)$	
AB	2041.89	1743.50	$\cot \alpha = 1.391\,932$
CB	-1493.36	976.64	$\cot \beta = 2.144\,643$
AC	3535.25	766.86	

Applied to the equation above, the bearing of PB is given by

$$\phi_{PB} = \tan^{-1}\left(\frac{406.3080}{986.1276}\right) = 22.3928°$$

and $\quad \phi_{BO} = \phi_{PB} + 180 \quad\quad = 202.3928°$

$$\begin{aligned}
-\alpha &= \underline{35.6944°} \\
\phi_{AO} &= 166.6984° \rightarrow \tan\phi_{AO} = -0.236\,420 \\
\phi_{BO} &= 202.3928° \\
+\beta &= \underline{24.9986°} \\
\phi_{CO} &= 227.3914° \quad \rightarrow \tan\phi_{CO} = \underline{1.087\,165} \\
& \quad\quad\quad\quad\quad\quad \tan\phi_{AO} - \tan\phi_{CO} = -1.323\,585
\end{aligned}$$

By equation 2.22,

$$\Delta N_{AO} = \frac{\Delta E_{AC} - \Delta N_{AC} \tan\phi_{CO}}{\tan\phi_{AO} - \tan\phi_{CO}}$$

and by equation 2.23,

$$\Delta E_{AO} = \Delta N_{AO} \tan\phi_{AO}$$

$\Delta N_{AO} = -2\,041.08$ m	$\Delta E_{AO} = 482.55$ m
$N_A = \underline{428\,373.70}$ m	$E_A = \underline{904\,365.74}$ m
$N_O = 426\,332.62$ m	$E_O = 904\,848.29$ m
Check	$\Delta E_{CO} = -3\,052.70$ m
	$E_C = \underline{907\,900.99}$ m
	$E_O = 904\,848.29$ m

Exercises 2.2

1 Stations X and Y are triangulation stations having the coordinates.

	E	N
X	27 465.312 m	64 313.765 m
Y	32 813.343 m	67 868.558 m

Two stations A and B, to the south of XY, are occupied and the angles observed are as follows.

$XAY = 77°\,10'\,24''$ $\quad YAB = 68°\,16'\,30''$
$ABX = 21°\,34'\,18''$ $\quad XBY = 73°\,32'\,06''$

Compute the coordinates of A and B.
(Ans. A: 29 690.607 m E, 62 301.819 m N   B: 31 507.371 m E,
62 062.516 m N)

2 In the quadrilateral $ABCD$, the length $AB = 1303.516$ m. The observed angles were as follows.

$ACB = 49°\,41'\,02''$ $\quad CDA = 45°\,37'\,45''$
$BCD = 48°\,07'45''$ $\quad ADB = 54°\,12'\,38''$

If D is to the east of C, compute the length of CD.

(Ans. 920.250 m)

3 The coordinates of three 'up' stations relative to a local origin are as follows.

	E	N
X	1011.10 m	3104.35 m
Y	3898.56 m	5838.71 m
Z	6833.74 m	4805.26 m

Observations taken from a station A, south of station Y, are as follows.

$XAY = 53°\,14'\,55''$ $\quad YAZ = 36°\,12'\,06''$

Calculate the coordinates of station A and the bearing of the line AY. (Ans. 3580.51 m E, 914.01 m N; $03°\,41'\,01''$)

4 The coordinates of three triangulation stations are given as follows.

	E (m)	N (m)
A	904 365.74	428 373.70
B	906 407.63	430 117.20
C	907 900.99	429 140.56

If the observed angles are $AOB = 315°\,41'\,40''$, $BOC = 304°59'\,55''$, compute the coordinates of the observation station O. (Ans. 906 415.54 m E, 430 287.14 m N)

5 The coordinates of four 'up' stations A, B, C and D are given as follows.

	E (m)	N (m)
A	440 747.87	398 539.17
B	441 865.65	404 304.23
C	446 777.18	404 641.86
D	450 906.16	401 810.40

Observations taken from two 'down' stations X and Y are given as follows.

$AXB = 83°\,28'\,20''$ $XYC = 123°\,12'\,45''$
$BXY = 81°\,18'\,57''$ $CYD = 75°\,38'\,54''$

Calculate the bearing of XY.   (Ans. $59°\,56'.16''$)

6 In order to fix the position of a borehole, observations were made from a station D to three triangulation 'up' stations A, B and C; the following clockwise angles were recorded: $ADB = 53°\,38'\,50''$ and $BDC = 61°\,39'\,40''$ (D is south of A, B and C). The Grid coordinates of the fixed stations are given as follows.

	E (m)	N (m)
A	436 572.06	374 821.17
B	439 248.29	375 664.70
C	441 627.81	373 893.93

(a) Determine whether the point D can be fixed without ambiguity
(b) State how the coordinates of D may be computed
(c) Compute the angles BAD and DCB.
           (RICS/M. Ans. $57°\,48'\,27''$, $61°\,02'\,05''$)

7 As part of a hydrographic survey, sextant readings were taken from a boat P offshore to three shore stations A, B and C and the recorded angles were $APB = 32°\,30'$ and $BPC = 62°\,30'$. The distances AB and BC are known to be 265.5 m and 324.8 m respectively and the angle ABC was $233°\,30'$ measured on the land side. Determine graphically the distance of the boat from B.

The boat is now moved inshore to P' and further readings taken as $AP'B = 99°\,30'$ and $BP'C = 113°\,30'$ (P' lies within the triangle ABC). Determine graphically the distance between the boat stations PP'. (RICS/M Ans. $BP = 368$ m, $PP' = 326$ m)

2.6.3 Semi-graphic resection

The basic principles are as in intersection; the differences are only in detail.

In this process, the trial point coordinates may be obtained by one of the following methods.

a) estimation from a small scale plot;
b) derivation by a graphical solution (Collins' point) using 3 rays;
c) computation by three point resection (Blunt's method).

The major difficulty with resection is that an orientation factor needs to be considered because there is no way in which the observer can obtain absolute bearings. The orientation factor may be largely eliminated if a Blunt solution is adopted and the remaining rays are then treated as an intersection. This is the method favoured here.

Example 2.7 Using the following data, compute the co-ordinates of the observers' station O.

	E(m)	N(m)	Assumed ϕ from O
A	40 747.87	48 539.17	314° 01′ 01.4″
B	41 865.65	54 304.23	043° 14′ 58.9″
C	46 777.18	54 641.86	161° 52′ 04.6″
D	50 906.16	51 810.40	199° 49′ 54.4″
E	43 804.81	49 349.76	279° 44′ 37.4″

Selecting rays B, C and E, then

$$\alpha = \phi_{OC} - \phi_{OB} = 161.8679° - 43.2497° = 118.6183°$$

$$\beta = \phi_{OE} - \phi_{OC} = 279.7437° - 161.8679° = 117.8758°$$

Solving by Blunt's method,

$$E'_O = 43\,788.26 \text{ m and } N'_O = 53\,043.75 \text{ m (Trial point)}$$

Using these coordinates the bearings from O' are given as follows.

$\phi^c_{OB} = 303.2492°$ $\qquad \phi^o_{OB} = 043.2497°$

(S) 2298.96(48) m

$\phi^c_{OC} = 061.8676°$ $\qquad \phi^o_{OC} = 161.8679°$

(S) 3389.33(60) m

$\phi^c_{OE} = 179.7433°$ $\qquad \phi^o_{OE} = 279.7437°$

(S) 3694.02(71) m

The orientation factor is then $\phi^c - \phi^o = -100.0004°$.

The other bearings are now reoriented as follows.

$\phi_{OA} = 214.0167°$ $\qquad \phi_{AO} = 34.0167°$ $\qquad S = 5434$ m

$\phi_{OD} = 099.8314°$ $\qquad \phi_{DO} = 279.8314°$ $\qquad S = 7224$ m

To compute the 'cut' coordinates, the reverse bearings are used, as in intersection.

Ray from station	\multicolumn{2}{c}{A}	\multicolumn{2}{c}{D}
Bearing from station	\multicolumn{2}{c}{034.0167°}	\multicolumn{2}{c}{279.8314°}
	E	N	E	N
Provisional coordinates	43 788.26	53 043.75	43 788.26	53 043.75
Station coordinates	40 747.87	48 539.17	50 906.16	51 810.40
$\Delta E / \Delta N$ (select larger)		4504.58	−7117.90	
$\cot\phi / \tan\phi$		0.67493	−0.17329	
dN/dE	3040.29			1233.49
'Cut' coordinates	43 788.16			53 043.89
Error		−0.10		0.14

2.6 RESECTION

The weights may be computed as follows.

	S	$w = 1/S^2 (\times 10^{-8})$	
AO	5430	3.39	(3)
BO	2300	18.90	(19)
CO	3390	8.70	(9)
DO	7220	1.92	(2)
EO	3690	7.34	(7)

Plotting the 'cut' coordinates produces the error graph shown in Fig. 2.12.

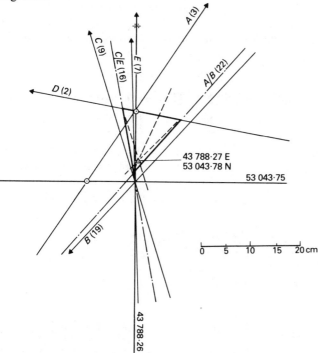

Fig. 2.12

2.6.4 Resection by 'variation of coordinates'

In resection, the bearings are reversed from those of intersection, i.e. ϕ_{AB} becomes ϕ_{BA}, and the observed bearings are subjected to an orientation factor (z). As

$$\phi_{BA} = \tan^{-1} \Delta E_{BA}/\Delta N_{BA}$$

then $\quad \delta\phi = \Delta N_{BA}(\delta E_B - \delta E_A)/S^2 - \Delta E_{BA}(\delta N_B - \delta N_A)/S^2$

As B is fixed, then

$$\delta\phi = -\Delta N_{BA}\delta E_B/S^2 + \Delta E_{BA}\delta N_B/S^2$$

(as in intersection).

The observational equations are then given in general terms as follows.

$$P_{oj}\delta E_o - Q_{oj}\delta N_o = (\phi^o_{oj} + r_i - z) - \phi^c_{oj} \qquad (2.24)$$

$$r_j = a_j\delta E_o + b_j\delta N_o + z - l_j \qquad (2.25)$$

where
$$a_j = P_{oj} = 206\,265\,\Delta N_{oj}/S^2_{oj}$$
$$= 206\,265\,\cos\phi_{oj}/S_{oj}$$
$$b_j = -Q_{oj} = -206\,265\,\Delta E_{oj}/S^2_{oj}$$
$$= -206\,265\,\sin\phi_{oj}/S_{oj}$$
$$l_j = \delta\phi = \phi^o - \phi^c$$
$$z = \text{the orientation factor}$$

The equations are normalised as below, and solved for δE, δN and z.

$$[aa]\delta E + [ab]\delta N + [a]z - [al] = 0$$
$$[ab]\delta E + [bb]\delta N + [b]z - [bl] = 0$$
$$[a]\delta E + [b]\delta N + mz - [l] = 0$$

As an alternative, z, which is common to all equations, may be eliminated as follows. Taking the normal equation in z as

$$mz = -\Big([a]\delta E + [b]\delta N - [l]\Big)$$

i.e.
$$z = -\left(\frac{[a]}{m}\delta E + \frac{[b]}{m}\delta N - \frac{[l]}{m}\right)$$

and if this is substituted back into the observational equations, then

$$r_j = a_j\delta E + b_j\delta N - \left(\frac{[a]}{m}\delta E + \frac{[b]}{m}\delta N - \frac{[l]}{m}\right) - l_i$$
$$= (a_i - [a]/m)\delta E + (b_j - [b]/m)\delta N - (l_j - [l]/m)$$
$$r_j = A_j\delta E + B_j\delta N - L_j \qquad (2.26)$$

The normal equations then become

$$[AA]\delta E + [AB]\delta N - [AL] = 0$$
$$[AB]\delta E + [BB]\delta N - [BL] = 0$$

which are then solved for δE and δN.

The following points should be noted.

1) $[A] = [B] = [L] = 0$
2) Here the angles are measured, and to produce bearings an orientated ray is assumed which is subjected to the orientation factor (z); thus all the assumed observed bearings are subjected to the same factor (z).

2.6 RESECTION

Example 2.8 Example 2.7 may be computed by the method of variation of coordinates, assuming coordinates for O of 43 788.26 m E, 53 043.75 m N.

The data may be tabulated as follows.

	ΔE_{ij}	ΔN_{ij}	$\phi°$	ϕ^c	$P_j = \Sigma\phi$	S_{ij}	$a = P_{oj}$	$b_j = -Q_{oj}$
AO	3040.39	4504.58	00.1	03.32	−3.22	5434.63	31.459	−21.233
BO	1922.61	−1260.48	57.1	57.12	−0.02	2298.97	−49.192	−75.033
CO	2988.92	−1598.11	03.4	03.22	0.18	3389.34	−28.695	53.667
DO	−7117.90	1233.35	46.1	49.03	−2.93	7223.96	4.875	28.134
EO	−16.55	3693.99	35.9	35.89	0.01	3694.03	55.837	0.250

Applying the matrix algebra as in intersection,

$$r = \begin{pmatrix} 31.459 & -21.233 & 1 \\ -49.192 & -75.033 & 1 \\ -28.695 & 53.667 & 1 \\ 4.875 & 28.134 & 1 \\ 55.837 & 0.250 & 1 \end{pmatrix} \begin{pmatrix} \delta E \\ \delta N \\ z \end{pmatrix} - \begin{pmatrix} -3.220 \\ -0.020 \\ 0.180 \\ -2.930 \\ 0.010 \end{pmatrix}$$

Solving as before,

$$x = (A^T A)^{-1} (A^T l)$$

$$(A^T A) = \begin{pmatrix} 7374.460 & 1634.192 & 14.284 \\ 1634.192 & 9752.522 & -14.215 \\ 14.284 & -14.215 & 5.000 \end{pmatrix}$$

$$(A^T A)^{-1} = \begin{pmatrix} 0.000142 & -0.000024 & -0.000475 \\ -0.000024 & 0.000107 & 0.000375 \\ -0.000475 & 0.000375 & 0.202422 \end{pmatrix}$$

$$(A^T l) = \begin{pmatrix} -119.205 \\ -2.889 \\ -5.980 \end{pmatrix}$$

Then $\quad x = \begin{pmatrix} \delta E \\ \delta N \\ z \end{pmatrix} = \begin{pmatrix} -0.0140 \\ 0.0004 \\ -1.1549 \end{pmatrix} \quad r = \begin{pmatrix} 1.6166 \\ -0.4733 \\ -0.9132 \\ 1.7171 \\ -1.9471 \end{pmatrix}$

$$r^T r = 10.411 \quad \sigma_0^2 = 5.205$$

$$\sigma_{xx} = \begin{pmatrix} 0.000739 & -0.000127 & -0.002473 \\ -0.000127 & 0.000558 & 0.001950 \\ -0.002473 & 0.001950 & 1.053693 \end{pmatrix}$$

$\sigma_E = \pm 0.027$ m $\sigma_N = \pm 0.024$ $\sigma_{E,N} = -0.000\,085$
$\sigma_z = \pm 1.026''$
$\sigma_{max} = 0.028$ m $\sigma_{min} = 0.023$ m $\phi_m = 114.0°$

The most probable coordinates are then $43\,788.25 + 0.03$ m E, $53\,043.75 + 0.02$ m N.

Alternative manual solution.

a	b	−l	A	B	L	Σ
31.459	−21.233	3.22	28.602	−18.390	2.024	12.236
−49.192	−75.033	0.02	−52.049	−72.190	−1.176	−125.415
−28.695	53.667	−0.18	−31.552	56.510	−1.376	23.582
4.875	28.134	2.93	2.018	30.977	1.734	34.729
55.837	0.250	−0.01	52.980	3.093	−1.206	54.867
14.284	−14.215	5.98	−0.001	0.000	0.000	−0.001

$[a]/m = 2.857$ $[b]/m = -2.843$ $[l]/m = 1.196$

The normal equations are then

$$7333.653\,\delta E + 1674.801\,\delta N + 102.121 = 0$$
$$1674.801\,\delta E + 9712.110\,\delta N + 19.900 = 0$$

Solving as before, $\delta E = -0.0140$, $\delta N = 0.0004$.

$$[rr] = -[AL]\delta E - [BL]\delta N + [LL] = 10.412$$

By equation 1.42,

$$\sigma_o^2 = [rr]/(m-n) = 10.412/3 = 3.471$$
$$\sigma_o = \pm 1.863$$

By equations 1.45 and 1.46,

$$\sigma_E = \sigma_o \sqrt{([BB]/K)} = \pm 0.03$$
$$\sigma_N = \sigma_o \sqrt{([AA]/K)} = \pm 0.02$$

Then $E_O = 43\,788.25 \pm 0.03$ m and $N_O = 53\,043.75 \pm 0.02$.

2.7 The direction method of triangulation

This method of computing 'by directions' is based upon semi-graphic processes and has been used since 1820; it has become popular for secondary and tertiary work throughout the world since its introduction in 1915 in South Africa by Dr W. C. van der Sterr and has become the standard method in many parts of the world, particularly for French Cadastre.

The principle advantages are as follows:

a) simplicity of computation—only basic semi-graphic principles are involved;

2.7 THE DIRECTION METHOD OF TRIANGULATION

b) flexibility in the field—less consideration of angular relationship;
c) simple analysis of the results by inspection.

The following points should be noted.

i) Here spherical excess is ignored to simplify the description but should not be forgotten for rays > 10 km.
ii) The method does not require any scale factor or true orientation until the final values of the station coordinates are required.

2.7.1 Definitions

A **direction** is similar to a bearing except that it represents the most probable bearing obtained between coordinated and uncoordinated points, based upon all the data available at the time of estimation. Directions increase in 'accuracy' as the work proceeds until stations are fixed, when the final bearings are fixed.

An observation made from a fixed station becomes a **forward direction**, whilst observations made from unfixed stations are **backward directions**. A mean of forward and backward directions is derived prior to intersection graphs being drawn.

A **fixed (datum) station** is one selected at the start of the system having fixed coordinates. The coordinates may be 'grid' or 'assumed'; the latter values would subsequently require orientation and the application of a scale factor to give the former.

Observational procedure is by conventional methods but sequence of station observations must be fairly rigidly followed and based upon the predetermined computational sequence. A ray is selected as the *RO* (reference object) having a known direction and the theodolite is orientated to this value within a few seconds of the given value.

The **booking system** allows the mean directions and not the angles to be extracted (see Fig. 2.14).

The **direction sheet** provides a method of deriving the best possible orientation of the rays as the work proceeds. The final direction can only come from the graph at the time the coordinates of the point being fixed are derived. Adjustment to the assumed bearings are computed and transferred to the direction sheet (see Fig. 2.15).

Coordinates of the station are by semi-graphic intersection techniques, and are based upon a minimum of five directions. Note that a doubly observed ray has two directions, i.e. forward and backward; a fully observed triangle has four fixing directions; and a singly observed ray takes half the weight of a doubly observed ray.

2.7.2 The direction method procedure

1) From a sketch of the triangulation layout, decide on the sequence of computation (and thus observation) to give the best fix from the initial datum stations. Note that 90° is the best 'cut' angle.
2) Compute the bearings between the datum stations.
3) Commence observing at the datum stations (orientating the theodolite) and observe all the rays that can be seen (see Fig. 2.14).
4) Transfer the mean directions on to the direction sheet (Fig. 2.15) and derive the mean directions of all the rays to be used in fixing the station.
5) Using two selected rays from the datum stations, compute, using 'intersection by bearings', the trial point coordinates.
6) Compute the 'cut' coordinates of the remaining rays.
7) Plot the intersection graph and solve the error figure to give the most probable coordinates of the point (see p. 54).
8) Derive adjustments to the assumed directions using the equation $\delta\phi'' = 206\,265\,p/S$. The sign of the adjustment depends upon the direction of the ray and the relative position to the most probable position of the fixed point.

Example 2.9 Consider the following datum stations.

	E (m)	N (m)
Plas	10 000.00	10 000.00
Castell	13 359.87	9 288.79
Bryn	10 537.12	12 971.21
Pimple	11 681.31	11 098.68

The stations to be fixed are: Igloo, East Base, and West Base (Fig. 2.13).

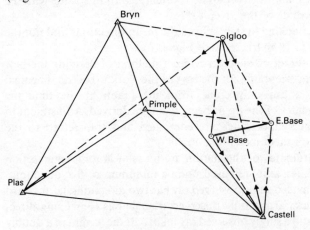

Fig. 2.13 Part of a triangulation scheme (not to scale)

2.7 THE DIRECTION METHOD OF TRIANGULATION

ROUND DIAGRAM   SHEET NO. 10

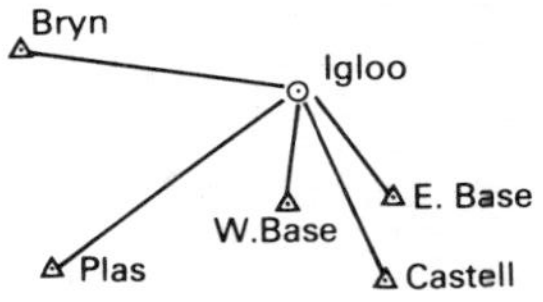

TRIANGULATION OBSERVATIONS AT STATION Igloo
Date 1.5.1980 Time 0930 Instrument Wild T2 Observer FAS
Weather Grey, overcast Booker FAS
Back bearing to station Plas 221° 17' 00"

Arc	Station		Plas	Bryn	E. Base	Castell	W. Base
1		L	221° 16' 54"	265° 35' 07"	165° 30' 20"	171° 08' 19"	182° 27' 32"
		R	16 58	35 05	30 24	08 17	27 29
	Mean direction		221 16 56	265 35 06	165 30 22	171 08 18	182 27 30
2		R	310 42 37	355 00 50	254 56 00	260 33 57	271 53 12
		L	42 39	00 44	56 03	34 01	53 09
	Mean Subtract		310 42 38	355 00 47	254 56 02	260 33 59	271 53 10
			89 25 42	89 25 42	89 25 42	89 25 42	89 25 42
	Mean direction		221 16 56	265 35 05	165 30 20	171 08 17	182 27 28
3		L	040 40 27	84 58 40	344 53 55	350 31 54	001 51 07
		R	40 30	58 37	53 51	31 52	51 02
	Mean add		040 40 28	84 58 38	344 53 53	350 31 53	001 51 04
			180 36 28	180 36 28	180 36 28	180 36 28	180 36 28
	Mean direction		221 16 56	265 35 06	165 30 21	171 08 21	182 27 32
4		R					
		L					
	Mean Subtract/add						
	Mean direction		221 16 56	265 35 04	165 30 17	171 08 24	182 27 32
	Accepted Direction		221° 16' 56"	265° 35' 05"	165° 30' 20"	171° 08' 20"	182° 27' 31"

Fig. 2.14

80 ORIENTATION

1		2			3	4	5	6	7	8	9		
		°	′	″	′ / ″						°	′	″
Plas													
Castell	△	101	57	22		−15					101	57	07
Bryn	△	010	15	08		−19					010	14	49
Igloo	△	041	17	11	16/55	(−16)							
Pimple	△	056	50	25		−13					056	50	12
						3)−47 = −16							
Castell													
Plas	△	281	57	20		−13					281	57	07
Pimple	△	317	09	36		−15					317	09	21
Bryn	△	322	31	58	32	−17					322	31	41
Igloo	△	351	08	47	14	(−15)							
E. Base	△	006	58	29		(−15)							
						3)−45 = −15							
Pimple													
Bryn	△	328	34	23		0					328	34	23
E. Base	△	113	07	58	54	(−4)							
Castell	△	137	09	26		−5					137	09	21
Plas	△	236	50	20		−8					236	50	12
						3)−13 = −4							
Bryn													
Plas	△	190	14	46		+3					190	14	49
Igloo	△	085	35	07	09	(+2)							
Castell	△	142	31	41		0			−2		142	31	41
Pimple	△	148	34	20		+3			+1		148	34	23
						3)6 = 2			+2				
Igloo													
Plas	△	221	16	56	17/01	−1	55	58		0	221	16	56
Bryn	△	265	35	05	10	+4	09	10		+6	265	35	11
E. Base	△	165	30	20	25	(+5)				+10			
Castell	△	171	08	20	25	+12	32	28		3)16 = +5	171	08	30
W. Base	△	182	27	31	36	(+5)							
						3)15 = +5							
E. Base													
Igloo	△	345	30	22	34	+3	25	30					
W. Base	△	278	13	51		+21							
Castell	△	186	57	53	58/05	2)24 = 12	58/14	58/09					

Fig. 2.15 Direction sheet

2.7 THE DIRECTION METHOD OF TRIANGULATION

The observation sheet is shown in Fig. 2.14 and the direction sheet in Fig. 2.15. The direction sheet (Fig. 2.15) may be explained as follows up to the derivation of the coordinates of Igloo.

1) Observation station in column 1 shown underlined.
2) Station observed in clockwise sequence—fixed stations are indicated by Δ.
3) Column 2. Recorded mean directions observed.
4) Column 9. Computed bearings between fixed stations are final directions.
5) Column 4. Provisional adjustment to observed values. (Note that the mean value is applied to forward direction of unfixed rays.)
6) For Igloo—orientated backward directions are brought down into column 5.
7) The provisional adjustment to the observed value is recorded in column 4—the mean value again applied to all rays to give values in column 3.
8) Column 6 gives the mean of the forward and backward rays. (Note that the triangular misclosure is thus adjusted.) These values are now used in the semi-graphic process.
9) Column 7. The correction from the graph is entered here.
10) The final bearings are obtained by applying (7) to (6) to give (9).
11) Column 8 is obtained by comparing (2) with (9). The mean value in column 8 is applied to any remaining ray observed.

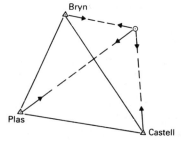

Fig. 2.16

The coordinates of Igloo may be fixed as follows (see Fig. 2.16). Bearings to Igloo:

    Plas   041° 16′ 58″
    Castell 351° 08′ 28″
    Bryn   085° 35′ 10″

The trial point coordinates are derived using Plas and Castell, using equations 2.1, 2.2, giving 12 759.21 E, 13 142.64 N. Finding the cut coordinates, using the process described on p. 52 for Bryn, gives 13 142.73, i.e. the error is +0.09, and ϕ = 085° 35′ 10″. The following table is then obtained, and the results may be transferred to the direction sheet.

Ray	$\simeq s$	$1/s^2$	W.	Adj. = $\delta\phi''$ =	206 265 p/s
Plas	4200	5.67×10^{-8}	6	$p = -0.044$	$\delta\phi'' = -2''$
Castell	3900	6.58×10^{-8}	7	$p = +0.038$	$\delta\phi'' = +2''$
Bryn	2200	2.07×10^{-7}	20	$p = +0.012$	$\delta\phi'' = +1''$

The coordinates of stations East Base and West Base are obtained in a similar manner.

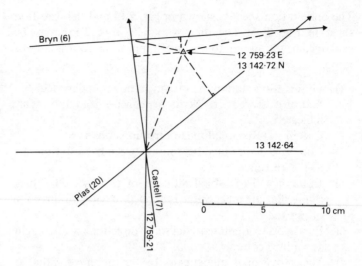

Fig. 2.17

The following points should be noted.

1) When all the points have been fixed the computed distance of the baseline is compared with the measured length of the base line to provide a scale factor K = (measured length/computed length).
2) By computing the true coordinates of a fixed station and deriving a true orientation the whole network may be transposed by equations 2.34, 2.35.

2.8 Satellite stations

If an inaccessible 'up' station is part of a triangulation, or if from a 'down' station observations to another station are impossible, a so-called 'satellite station' may be created where the required observations are possible. It is then necessary to reduce the measured values to the values which would have been measured had the initial station been occupied.

In Fig. 2.18, A and B are the fixed triangulation stations, C the inaccessible 'up' station and S the satellite station. The quantity θ_a represents the difference in direction between CA and SA, and θ_b represents the difference in direction between CB and SB. In $\triangle ASC$ the angle CSA (ϕ_a) is measured, then

$$\sin \theta_a = (CS \sin \phi_a)/AC$$

If θ_a is small then $\theta_a'' = 206\,265\, CS \sin \phi_a/AC$, and similarly $\theta_b'' = 206\,265\, CS \sin \phi_b/BC$. If the line SC is regarded as the arbitrary meridian then ϕ_a and ϕ_b represent arbitrary bearings of the lines SA and SB and thus the general equation becomes

$$\theta'' = \frac{K \sin \phi}{d} \qquad (2.27)$$

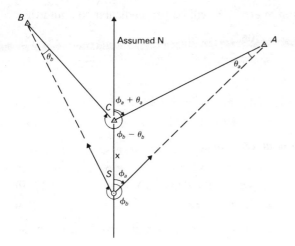

Fig. 2.18

where ϕ = the adjustment to the arbitrary bearing
$K = 206\,265\,X$
X = the length CS
d = the length of the ray

Note that, the sign of the adjustment θ is dependent on sin ϕ, i.e. if $\phi < 180°$, θ +ve; if $\phi > 180°$, θ −ve.

2.8.1 The effects of an error in θ by accepting the approximation equation

If $\quad \sin\theta = X \sin\phi/d \quad$ and

$$\sin\theta = \theta - \frac{\theta^3}{3!} + \frac{\theta^5}{5!} - \frac{\theta^7}{7!} \quad \text{etc.} \quad \text{(See Appendix 3)}$$

then $\quad \theta = (X \sin\phi/d) - (X \sin\phi/d)^3/6 \quad$ etc.

In order that the approximation is acceptable, the second and subsequent terms must be neglected and the limit is when

$$(X \sin\phi/d)^3/6 > \delta\theta$$

For maximum error, $\sin\phi = 1$, i.e. when $\phi = n\pi/2$. If $\delta\theta = 1''$ then $X^3/6d^3 = 1/206\,265$ and the error ratio is $X/d = 1/32.5$. For example, if $d = 10\,\text{km}$ then $X = 10\,000/32.5 \simeq 300\,\text{m}$. To improve the accuracy it can be seen that X should be as small as possible.

The effect of an error in X
From the basic equation $\theta = X \sin\phi/d$, by the application of partial differentiation

$$\delta\theta_X = \sin\phi \, \delta X/d = \theta \, \delta X/X \qquad (2.28)$$

For a maximum error, X will be perpendicular to d and then

$$\delta\theta''_{X(\max)} = 206\,265\,\delta X/d \tag{2.28a}$$

The effect of an error in d
Similarly

$$\delta\theta_d = -X\sin\phi\,\delta d/d^2 = -\theta\,\delta d/d \tag{2.29}$$

$$\delta\theta''_{d(\max)} = -206\,265\,X\,\delta d/d^2 \tag{2.29a}$$

The effect of an error in ϕ
Similarly

$$\delta\theta_\phi = X\cos\phi\,\delta\phi/d = \theta\tan\phi\,\delta\phi \tag{2.30}$$

$$\delta\theta''_{\phi(\max)} = 206\,265\,X\,\delta\phi/d \tag{2.30a}$$

From the above equations, the maximum error ratios are the same, i.e.

$$\delta X/X = \delta d/d = \delta\theta/\theta = d\,\delta\theta''/206\,265\,X = d\,\delta\theta''/K \tag{2.31}$$

Example 2.10 Given $X = 10\,\text{m}$, $d = 10\,\text{km}$, $\delta\theta = \pm 1''$, find the error ratio in θ.

As $\quad \delta\theta''_x = 206\,265\,\delta X/d$ then

$$\delta X = d\,\delta\theta''/206\,265 = 10\,000/206\,265 = \pm 0.048\,\text{m}$$

$$\delta X/X = 0.048/10 \simeq 1/200$$

As $\quad \delta\theta''_d = -206\,265\,X\,\delta d/d^2$ then

$$\delta d = d^2\,\delta\theta''/206\,265\,X = 10\,000^2/2\,062\,650 = \pm 50\,\text{m}$$

$$\delta d/d = 50/10\,000 \simeq 1/200$$

As $\quad \delta\theta''_\phi = X\,\delta\phi/d$ then

$$\delta\phi'' = d\,\delta\theta''/X = 10\,000/10 = \pm 1000'' = 16'\,40''$$

As $\quad \theta'' = 206\,265\,X/d = 206\,265/1000 = 206''$ then

$$\delta\theta/\theta \simeq 1/200$$

The following points can be seen from the above.

1) X must be measured accurately as it is required to be as small as is practicable.
2) d need only be known approximately as the error ratio is constant.
3) The bearings need only be derived to the same accuracy as the triangulation but orientation need only be approximate.
4) The error ratio is maximum and constant when $\delta\theta/\theta = \delta X/X = \delta d/d = d\,\delta\theta/206\,265X$.

Example 2.11 Given that the known bearing of the line CA is $172°\,26'\,32''$, compute the bearings of the other rays from the following recorded data.

Station set at S.
Observed relative bearings SA 000° 00′ 06″
                            SB 046° 57′ 36″
                            SC 092° 14′ 00″
                            SD 163° 26′ 36″

Approximate lengths 2.5×10^4 m
                     3.2×10^4 m
                     8.6 m
                     1.7×10^4 m

Tabulating (see Fig. 2.19),

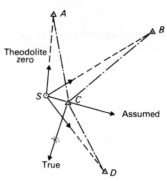

Fig. 2.19

Stn	Observed ϕ	Reduced ϕ	$\sin \phi_{red}$	Distances (m)	θ''	Adjusted ϕ	Reorientated ϕ
A	000° 00′ 06″	267° 46′ 06″	−0.999 24	2.5×10^4	−71	267° 44′ 55″	172° 26′ 32″
B	046° 57′ 36″	314° 43′ 36″	−0.710 39	3.2×10^4	−39	314° 42′ 57″	219° 24′ 34″
C	092° 14′ 00″	000° 00′ 00″		8.6		000° 00′ 00″	264° 41′ 37″
D	163° 26′ 36″	071° 12′ 36″	0.946 71	1.7×10^4	99	071° 14′ 15″	335° 55′ 52″

Exercises 2.3

1 In a quadrilateral *ABCD*, in clockwise order, forming part of a triangulation, a church spire was observed as the central station *O*. Accordingly, a satellite station *S* was selected 6.60 m from *O*, and inside the triangle *BOC*. The following table gives the approximate distances from the central station and the angles observed from *S*.

Observed station	Horizontal angle at S measured clockwise from O	Distance (m) from O
A	30° 45′ 30″	5445
B	98° 32′ 00″	6682
C	210° 10′ 40″	3858
D	320° 14′ 15″	4596
O	360° 00′ 00″	

Calculate the four angles at *O*.
(LU Ans. 67° 47′ 44″, 111° 32′ 21″, 110° 03′ 23″, 70° 36′ 32″)

2 In joining the triangulation of a survey for a new satellite town to the existing system, angles were measured from two stations *A* and *B* in the old triangulation to a flagstaff *C* on a church tower lying in the new area. To complete the adjustment of the triangulation, it is necessary to know the angle at *C* between the lines *AC* and *BC*. A station was set up at *S*, 10.21 m from the flagstaff and the following angles measured.

Station	Pointing	Angle
S	A	0° 00′ 00″
	B	59° 29′ 40″
	C	131° 53′ 00″

If the lengths of AC and BC were 3530 m and 8700 m respectively compute the angle ACB.　　(LU Ans. 59° 33′ 13″)

3　A, B and C are triangulation 'up' stations. It is required to find the coordinates of a station D on the top of a colliery headgear which cannot be occupied. From a satellite station S arbitrary bearings were recorded together with other relevant data as follows.

	E (m)	N (m)	Assumed bearing from S	Approximate distance from S (m)
A	381 462.36	449 271.47	352° 20′ 04″	5500
B	387 148.72	450 418.72	036° 48′ 16″	8260
C	383 175.68	438 716.72	169° 08′ 45″	5200
D	382 200	443 800 (approx)	0° 01′ 15″	10.365

From the data given calculate the most probable value of the coordinates of D.
　　　　(RICS/M Ans. 382 199.68 m E, 443 811.90 m N)

4　(a) Discuss the use of 'satellite' stations in triangulation observations stating the relative accuracies of the various measurements taken and the eccentric distance of the satellite station relative to the lengths of the rays.
(b) A, B, C and D are minor triangulation stations but the angles at C cannot be obtained due to an obstruction. A satellite station X has been set out to the west of C and the following information obtained.

Azimuth	XA	063° 15′ 26″	Distance	CX 4.682 m
	XC	094° 12′ 32″		CA 1816 m
	XB	116° 43′ 35″		CB 1320 m
	XD	254° 51′ 18″		CD 1500 m

Calculate the true bearings of CA, CB and CD.
　　　　(RICS/M Ans. 63° 10′ 52″, 116° 48′ 15″, 254° 54′ 51″)

2.9 The crossed quadrilateral

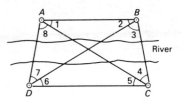

For the control of certain structures it may be necessary to set up four control stations which are totally intervisible (see Fig. 2.20).

Let the observed angles be numbered cyclically as shown.

Fig. 2.20

2.9.1 The angle conditions

The angle conditions are as follows (see Fig. 2.20).

(a) $(1) + (2) + (3) + (4) + (5) + (6) + (7) + (8) = 360°$
(b) $(1) + (2) - (5) - (6) = 0°$
(c) $ (3) + (4) - (7) - (8) = 0°$
(d) $(1) + (2) + (3) + (4) = 180°$
(e) $ (3) + (4) + (5) + (6) = 180°$
(f) $ (5) + (6) + (7) + (8) = 180°$
(g) $(1) + (2) + (7) + (8) = 180°$

It can be seen that $(d) - (e) = (b)$
$(e) - (f) = (c)$
$(e) + (g) = (a)$

Therefore equations (d), (e), (f) and (g) are redundant.

2.9.2 The side equations

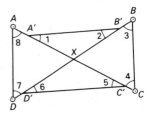

In order that the sides are compatible with the angles a further equation is necessary. Fig. 2.21 shows that the above angle conditions could be satisfied although the shape is not compatible.

In Fig. 2.21,

Fig. 2.21

$$BX = AX \sin 1 / \sin (2) \quad \text{if } AA' = BB' = 0$$
$$CX = BX \sin (3) / \sin (4) \quad \text{if } CC' = 0$$
$$DX = CX \sin (5) / \sin (6) \quad \text{if } DD' = 0$$
$$AX = DX \sin (7) / \sin (8)$$
$$= \frac{AX \sin (1) \sin (3) \sin (5) \sin (7)}{\sin (2) \sin (4) \sin (6) \sin (8)}$$

Therefore

$$\frac{\sin (1) \sin (3) \sin (5) \sin (7)}{\sin (2) \sin (4) \sin (6) \sin (8)} = 1 \qquad (2.32)$$

This is known as the **side equation** with X as the pole.

In linear form the equation becomes

$$\log \sin (1) + \log \sin (3) + \log \sin (5) + \log \sin (7)$$
$$- \log \sin (2) - \log \sin (4) - \log \sin (6) - \log \sin (8) = 0$$
(2.32a)

2.9.3 Non-rigorous adjustment of a braced quadrilateral by 'equal shifts'

The angles are adjusted by equal amounts at three separate stages and the side equation is then applied to the adjusted angles as follows.

1) Angles (1) to (8) are adjusted equally to sum to 360.
2) The sum of angles (1) and (2) are adjusted to equal the sum of angles (5) + (6).
3) The sum of angles (3) and (4) are adjusted to equal the sum of angles (7) + (8).
4) The side equation is applied by adjusting the angles so that the sum of the log sin 'odd' angles equals the sum of the log sin 'even' angles as follows.

Angle no.	Observed angle	A_1	Adj. angle	Pair	A_2	Adj. angle
1	28° 36′ 49″	0.25″	49.25″		−3.00″	46.25″
2	50° 08′ 26″	0.25″	26.25″	75.50″	−3.00″	23.25″
5	47° 43′ 05″	0.25″	05.25″		3.00″	08.25″
6	31° 01′ 58″	0.25″	58.25″	63.50″	3.00″	61.25″
3	44° 18′ 09″	0.25″	09.25″		3.50″	12.75″
4	56° 56′ 34″	0.25″	34.25″	43.59″	3.50″	37.75″
7	60° 40′ 19″	0.25″	19.25″		−3.50″	15.75″
8	40° 34′ 38″	0.25″	38.25″	57.50″	−3.50″	34.75″
	359° 59′ 58″	2.00″	00.00″			00.00″

Side equation

	Log sines		Δ/″ (×10⁻⁷)	A_3	Final adjusted angles	
1	9.680 2346		38.6	−0.5″	28° 36′ 45.75″	(45.8″)
2		9.885 1409	17.6	0.5″	50° 08′ 23.75″	(23.8″)
5	9.869 1459		19.1	−0.5″	47° 43′ 07.75″	(07.8″)
6		9.712 2639	35.0	0.5″	31° 02′ 01.75″	(01.8″)
3	9.844 1412		21.6	−0.5″	44° 18′ 12.25″	(12.3″)
4		9.923 3146	13.7	0.5″	56° 56′ 38.25″	(38.3″)
7	9.940 4277		11.9	−0.5″	60° 40′ 15.25″	(15.3″)
8		9.813 2208	24.6	0.5″	40° 34′ 35.25″	(35.3″)
	39.333 9494	39.333 9402	182.1		360° 00′ 00.00″	
	9402					
	92					

a) The log sin of each angle is recorded together with its difference per second.
b) the adjustment per second to each angle is given by the equation

$$\text{Adjustment} = \frac{\text{difference in } \sum \log \sin}{\sum \text{difference per second}} \quad \text{(in seconds)} \tag{2.33}$$

Thus $\Delta\Sigma - 92 \times 10^{-7}$

Final adjustment $A_3 = \Delta\Sigma \log \sin / \Sigma\Delta$ per sec
$= 92/182 = 0.5''$

2.9.4 Rigorous adjustment

Conditional equations are set up as follows.

$$\left.\begin{array}{l} r_1 + r_2 + r_3 + r_4 + r_5 + r_6 + r_7 + r_8 - X_1 = 0 \\ r_1 + r_2 \qquad\qquad -r_5 - r_6 \qquad\qquad -X_2 = 0 \\ \qquad\qquad r_3 + r_4 \qquad\qquad -r_7 - r_8 - X_3 = 0 \end{array}\right\} \begin{array}{l} \text{Angle} \\ \text{equations} \end{array}$$

$r_1\Delta_1 - r_2\Delta_2 + r_3\Delta_3 - r_4\Delta_4 + r_5\Delta_5 - r_6\Delta_6 + r_7\Delta_7 - r_8\Delta_8 - Y$
$\qquad\qquad = 0 \quad \text{Side equation}$

where X_1, X_2 and X_3 are the absolute values (see p. 32), Δ_1, $\Delta_2, \ldots, \Delta_8$ are the log sin (diff/sec) and $Y = \Sigma$ log sin odds $-\Sigma$ log sin evens. A correlative normal equation solution is applied and the eight simultaneous equations are solved. Applied to the previous example, the conditional equations are

$r_1 + r_2 + r_3 + r_4 + r_5 + r_6 + r_7 + r_8 \qquad -02.0'' = 0$
$r_1 + r_2 \qquad\qquad -r_5 - r_6 \qquad\qquad +12.0'' = 0$
$\qquad r_3 + r_4 \qquad\qquad -r_7 - r_8 \qquad -14.0'' = 0$
$38.6 r_1 - 17.6 r_2 + 21.6 r_3 - 13.7 r_4 + 19.1 r_5 - 35.0 r_6$
$\qquad\qquad + 11.9 r_7 - 24.6 r_8 + 92 = 0$

Deriving the normal equations and solving these gives

$r_1 = -3.38'' \quad r_2 = -2.17'' \quad r_3 = 3.40''$
$r_4 = 4.16'' \quad r_5 = 2.64'' \quad r_6 = 3.80''$
$r_7 = -3.62'' \quad r_8 = -2.83''$

Their sum = 2.00 (Check).
The angles are then

(1) 28° 36′ 45.6″   (2) 50° 08′ 23.8″
(3) 44° 18′ 12.4″   (4) 56° 56′ 38.2″
(5) 47° 43′ 07.6″   (6) 31° 02′ 01.8″
(7) 60° 40′ 15.4″   (8) 40° 34′ 35.2″

It can be seen that the maximum discrepancy is 0.2" and thus the rigorous adjustment is seldom justified compared with the considerable amount of extra work involved. In larger triangulation networks the variation of coordinates method is now generally preferred.

2.10 Correlation of underground and surface surveys

There are three main methods of transferring bearings and coordinates from a surface base line to an underground base line when the only access is via a vertical shaft(s):

1) using one wire in each of two shafts;
2) using two wires in a single shaft by the Weisbach triangle method;
3) by the use of the gyro-theodolite.

2.10.1 One wire in each of two shafts

Generally the coordinates of the wires in the shafts are known as (E_x, N_x) and (E_y, N_y), resection or intersection processes (see p. 47) having been used to connect the wires at the surface to the grid. The wires are assumed to lie in a vertical plane and thus the coordinates underground are assumed identical with those at the surface (Fig. 2.22). An underground traverse is made, based upon an assumed meridian, and the traverse coordinates are computed. A comparison between the length and bearing XY may be made and the grid transformation process applied.

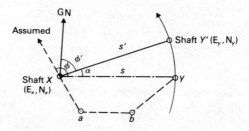

Fig. 2.22

The general equations are as follows.

$$E'_a = E'_x + k(E_a \cos\alpha - N_a \sin\alpha)$$
$$= E'_x + mE_a - nN_a \qquad (2.34)$$

and $$N'_a = N'_x + k(N_a \cos\alpha + E_a \sin\alpha)$$
$$= N'_x + mN_a + nE_a \qquad (2.35)$$

2.10 CORRELATION OF UNDERGROUND AND SURFACE SURVEYS

where

E'_a, N'_a = the transformed coordinates of the traverse coordinates (E_a, N_a)

E'_x, N'_x = the point of swing (coordinates O, O on the traverse)

$m = k \cos \alpha \quad n = k \sin \alpha$

k = the scale factor XY'/XY

α = the angle of swing = $\tan^{-1} n/m = \phi_{XY} - \phi_{XY'}$

In matrix form this is given as

$$\begin{pmatrix} E' \\ N' \end{pmatrix} = \begin{pmatrix} E'_0 \\ N'_0 \end{pmatrix} + \begin{pmatrix} m & -n \\ n & m \end{pmatrix} \begin{pmatrix} E \\ N \end{pmatrix} \qquad (2.36)$$

Example 2.12 During a correlation survey between a wire in each of two shafts, A and E, the following partial coordinates were obtained.

Line	ΔE (ft)	ΔN (ft)
AB	+392.593	−29.873
BC	+260.820	+45.794
CD	+295.125	+158.125
DE	−44.112	+103.488

After resection, the grid coordinates of A and E were found to be as follows.

	E (m)	N (m)
A	556 821.63	447 219.42
E	557 098.39	447 300.38

Transform the partial coordinates into the grid coordinates for each station. (RICS/M)

From the partial coordinates of the traverse the assumed coordinates are given as follows.

	E (ft)	N (ft)
A	0.000	0.000
B	392.593	−29.873
C	653.413	15.921
D	948.538	174.046
E	904.426	277.534

Traverse values:

$$\phi_{AE} = \tan^{-1} 904.426/277.534 = 72.9407°$$
$$S_{AE} = (904.426^2 + 277.534^2)^{1/2} = 946.0505 \text{ ft}$$

Grid values:

$$\phi'_{AE} = \tan^{-1} 276.76/80.96 = 73.6943°$$
$$S'_{AE} = (276.76^2 + 80.96^2)^{1/2} = 288.3585 \text{ m}$$

Then

$$\text{swing} = \alpha = \phi_{AE} - \phi'_{AE} = 72.9407° - 73.6943°$$
$$= -0.7536°$$

Scale factor $= k = S'/S = 288.3585/946.0505 = 0.304\,8024$

$$m = k \cos \alpha = 0.304\,776\,036$$
$$n = k \sin \alpha = -0.004\,008\,890$$
$$E'_O = E'_A = 556\,821.63 \qquad N'_O = N'_A = 447\,219.42$$
$$E'_P = E'_A + mE_P - nN_P \quad \text{and} \quad N'_P = N'_A + mN_P + nE_P$$

Then

$$E'_B = 556\,821.63 + (392.593m) - (-29.873n) = 556\,941.16 \text{ m}$$

similarly

$$E'_C = \qquad\qquad\qquad\qquad\qquad = 557\,020.84 \text{ m}$$
$$E'_D = \qquad\qquad\qquad\qquad\qquad = 557\,111.42 \text{ m}$$
$$E'_E = 556\,821.63 + (904.426m) - (277.534n)$$
$$= 557\,098.39 \text{ m} \quad \text{(check)}$$
$$N'_B = 447\,219.42 + (-29.873m) + (392.593n) = 447\,208.74 \text{ m}$$

similarly

$$N'_C = 447\,221.65 \text{ m}$$
$$N'_D = 447\,268.66 \text{ m}$$
$$N'_E = 447\,219.42 + (277.534m) + (904.426n)$$
$$= 447\,300.38 \text{ m} \quad \text{(check)}$$

2.10.2 Two wires in a single shaft (Weisbach triangle)

There are several alternative methods but only the most favoured one is illustrated.

Fig. 2.23 shows a typical layout. At the surface, the base line PQ has been connected to grid and thus the coordinates of $P(E_P, N_P)$ and $Q(E_Q, N_Q)$ are known. A theodolite is set on the surface at P and many observations (say 16) are made of angles $Q\hat{P}B$ and $Q\hat{P}A$. These will not be consistent as the wires are likely to oscillate. The mean value is $\alpha_s = Q\hat{P}A - Q\hat{P}B$. In the Weisbach triangle APB, P is approximately aligned with the wires A and B and thus α is small (less than 20'). The distances $PA = b$, $PB = a$ and $AB = p$ are measured. The distance b

2.10 CORRELATION OF UNDERGROUND AND SURFACE SURVEYS

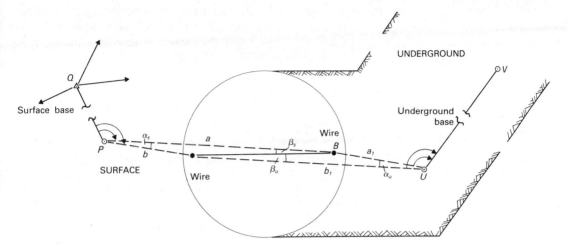

Fig. 2.23

should be less than or equal to p. To obtain the bearing of the plumb-plane (ϕ_{AB}), the angle β_s must be computed.

$$\sin \beta_s = b \sin \alpha_s / p \quad (2.37)$$

but as α and β are very small then

$$\beta_s'' = b \alpha_s'' / p \quad (2.37a)$$

$$\phi_{BA} = \phi_{PQ} + Q\hat{P}B \pm 180 - \beta_s = \phi_{QP} + Q\hat{P}B - \beta_s$$

Underground the theodolite station is U and the underground base line is UV. Angles AUV and BUV are measured and then

$$\alpha_u = A\hat{U}V - B\hat{U}V$$
$$\beta_U'' = a_1 \alpha_U'' / p$$

and $\quad \phi_{UV} = \phi_{AB} + \beta_U \pm 180 + A\hat{U}V = \phi_{BA} + \beta_U + A\hat{U}V$

Theory of the Weisbach triangle
In Fig. 2.24, $\sin \beta = b \sin \alpha / c$

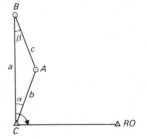

Fig. 2.24

Successive partial differentiation with respect to the separate variables gives the errors ($\delta \beta$) (see p. 12).

$$\cos \beta \, \delta \beta_b = \sin \alpha \, \delta b / c = \sin \beta \, \delta b / b$$
$$\delta \beta_b = \tan \beta \, \delta b / b$$

similarly

$$\delta \beta_\alpha = \tan \beta \, \delta \alpha / \tan \alpha$$
and $\quad \delta \beta_c = \tan \beta \, \delta c / c$

Then the rms value is

$$\delta \beta = \tan \beta [(\delta b / b)^2 + (\delta c / c)^2 + (\cot \alpha \, \delta \alpha)^2]^{1/2} \quad (2.38)$$

The first two terms under the radical sign represent the linear error ratios, whilst the third term represents the effect on β due to the angular error $\delta\alpha$. The effect of linear errors will be negligible compared with the angular error if their ratio is 1:3 (criteria of negligibility p. 13), i.e.

$$\delta\alpha^2/\tan^2\alpha \geqslant 9(c^2\delta b^2 + b^2\delta c^2)/b^2 c^2$$
$$\tan\alpha \leqslant bc\delta\alpha/[3\sqrt{(c^2\delta b^2 + b^2\delta c^2)}] \qquad (2.39)$$

The error in α, i.e. $\delta\alpha$, will be a minimum when $\tan\alpha = 0$, i.e. when $\alpha = 0$. Thus α should be as small as possible and the transmitted error $\delta\beta$ is diminished when $c > b$.

Example 2.13 Let $c = b = 3.0000 \pm 0.0025$ m; $\delta\alpha = \pm 5''$. Then

$$\tan\alpha \leqslant (bc\,\delta\alpha)/[3\sqrt{(c^2\delta b^2 + b^2\delta c^2)}]$$

As α is small,

$$\alpha'' \leqslant (bc\,\delta\alpha'')/[3\sqrt{(c^2\delta b^2 + b^2\delta c^2)}]$$
$$\leqslant (9.0000 \times \pm 5'')/[3\sqrt{(9.0000 \times 6.25 \times 10^{-6} \times 2)}]$$
$$\leqslant 15/\pm 0.0106 \leqslant \pm 1414'' = \pm 23.5'$$

Note that when $\alpha \leqslant 23.5'$, $\tan\alpha \simeq \alpha_{\text{rad}} \simeq \sin\alpha$ to 6 places of decimals.

$$\tan 23.5' = 0.006\,835\,979$$
$$23.5'_{\text{rad}} = 0.006\,835\,873$$
$$\sin 23.5' = 0.006\,835\,820$$

The basic equation is now acceptable as $\beta = b\alpha/c$. Then

$$\delta\beta_b = \alpha\,\delta b/c = \beta\,\delta b/b$$
$$\delta\beta_c = -b\alpha\,\delta c/c^2 = -\beta\,\delta c/c$$
and $\quad \delta\beta_\alpha = b\alpha/c = \beta\,\delta\alpha/\alpha$

The rms value is then

$$\delta\beta = \beta[(\delta b/b)^2 + (\delta c/c)^2 + (\delta\alpha/\alpha)^2]^{1/2} \qquad (2.40)$$

Example 2.14 Given $b = 3.000 \pm 0.005$ m, $c = 6.000 \pm 0.005$ m, $\alpha = 0°\,10'\,00'' \pm 5''$. Then

$$\beta'' = b\alpha''/c = 3 \times 600''/6 = 300'' = 0°\,05'\,00''$$
$$\delta\beta_b = \beta\,\delta b/b = 300'' \times \pm 0.005/3.000 = \pm 0.50''$$
$$\delta\beta_c = -\beta\,\delta c/c = -300'' \times \pm 0.005/6.000 = \pm 0.25''$$
$$\delta\beta_\alpha = \beta\,\delta\alpha/\alpha = 300'' \times \pm 5''/600'' = \pm 2.50''$$
$$\delta\beta = (0.50^2 + 0.25^2 + 2.50^2)^{1/2} = \pm 2.56''$$

It can be seen from the above that:

1) the linear values need only be measured with normal care;

2.10 CORRELATION OF UNDERGROUND AND SURFACE SURVEYS

2) the effect of the error in α is distributed to the angle β by a multiplying factor b/c. If $b < c$ the error in $\beta <$ the error in α.

Example 2.15 The following observations were made as part of a correlation survey using the Weisbach triangle method. At the surface, the clockwise angles were made to a reference line AB, which had a National Grid bearing of $208°\,08'\,20''$.

Observed values: $AB\,Y = 92°\,22'\,26''$ $BY = 2.42\,\text{m}$
$ABX = 92°\,35'\,38''$ $YX = 3.18\,\text{m}$

Underground clockwise angles were made from a reference line CD which lies to the south of the wires X and Y.

$DCY = 170°\,56'\,30''$
$DCX = 170°\,58'\,22''$; $CX = 2.84\,\text{m}$.

Assuming that all the angles were measured with a standard error of $\pm 3''$ and the lengths $\pm 0.01\,\text{m}$, calculate the most probable value of the bearing of the underground reference line CD and its standard error.   (RICS/M)

Fig. 2.25

In Fig. 2.25,

$$\alpha_s = A\hat{B}X - A\hat{B}Y$$
$$= 92°\,35'\,38'' - 92°\,22'\,26''$$
$$= 0°\,13'\,12'' = 792''$$
$$\beta_s'' = \alpha_s'' x/b = 792'' \times 2.42/3.18 = 603'' = 10'\,03''$$
$$\alpha_u = D\hat{C}X - D\hat{C}Y$$
$$= 170°\,58'\,22'' - 170°\,56'\,30''$$
$$= 0°\,01'\,52'' = 112''$$
$$\beta_u'' = \alpha_u'' y/b = 112 \times 2.84/3.18 = 100'' = 01'\,40''$$
$$\phi_{XY} = \phi_{BA} + A\hat{B}X + \beta_s \pm 180$$
$$= 28°\,08'\,20'' + 92°\,35'\,38'' + 0°\,10'\,03'' + 180°$$
$$= 300°\,54'\,01''$$
$$\phi_{CD} = \phi_{YX} + \beta_u - D\hat{C}Y \pm 180$$
$$= 120°\,54'\,01'' + 0°\,01'\,40'' - 170°\,56'\,30'' + 180°$$
$$= 129°\,59'\,11''$$

If the error in $A\hat{B}X = \pm 3''$ and in $A\hat{B}Y = \pm 3''$, then the error in α_s is

$$\delta\alpha_s = \delta\alpha_u = \sqrt{2} \times 3'' = \pm 4.24''$$
$$\phi_{CD} \pm \delta\phi_{CD} = \phi_{BA} + (A\hat{B}X \pm 3'') + (\beta_s \pm \delta\beta_s)$$
$$+ (\beta_u \pm \delta\beta_u) - (D\hat{C}Y \pm 3'')$$
$$\delta\phi_{CD} = (3^2 + \delta\beta_s^2 + \delta\beta_u^2 + 3^2)^{1/2}$$

By equation 2.40,

$$\delta\beta_s = \beta_s''[(\delta x/x)^2 + (\delta b/b)^2 + (\delta\alpha/\alpha)^2]^{1/2}$$
$$= 603[(0.01/2.42)^2 + (0.01/3.18)^2 + (4.24/792)^2]^{1/2}$$
$$= \pm 4.5''$$

Similarly

$$\delta\beta_u = \beta_u''[(\delta y/y)^2 + (\delta b/b)^2 + (\delta\alpha/\alpha)^2]^{1/2}$$
$$= 100[(0.01/2.84)^2 + (0.01/3.18)^2 + (4.24/112)^2]^{1/2}$$
$$= \pm 3.8''$$
$$\delta\phi_{CD} = (3.0^2 + 4.5^2 + 3.8^2 + 3.0^2)^{1/2}$$
$$= \pm 7.26''$$

The most probable value of the bearing ϕ_{CD} is $129°\,59'\,11''$ $\pm 7.3''$.

Exercises 2.4

1 Discuss the method of correlation using the 'Weisbach triangle' stating its advantages and disadvantages and giving the ideal conditions for such a system. (RICS/M)

2 The following observations were made as part of a correlation using the 'Weisbach Triangle' method. At the surface, the clockwise angles were made to a base line AB, which has a bearing $268°\,18'\,20''$.

$AB\,Y = 102°\,32'\,26''$ $BY = 2.52$ m
$ABX = 102°\,45'\,38''$ $YX = 3.20$ m

Underground, clockwise angles were then made to a reference line CD, which lies to the south of the wires.

$DC\,Y = 176°\,58'\,33''$
$DCX = 176°\,56'\,18''$ $CX = 2.96$ m

Assuming that the angles were measured with a standard error of $\pm 3''$ and the lengths ± 0.01 m, calculate the most probable value of the bearing of the underground reference line CD.
(RICS/M Ans. $194°\,13'\,43.8'' \pm 7.4''$)

3 Explain the use of the 'Weisbach triangle' for setting out underground. The centre line of a tunnel is represented by two plumb-lines, A and B, 4.100 m apart, hanging vertically in a shaft, the whole circle bearing of the line AB being $80°\,40'\,15''$. A theodolite is set up underground at C distant 3.515 m and roughly east of the nearer plumb line B, and the observed value of the angle ACB is found to be $16'\,12''$. Calculate the bearing of the line AC and the perpendicular distance of C from the centre line of the tunnel.   (LU Ans. $80°\,54'\,08''$, 0.031 m)

4 Two plumb wires *A* and *B* in a shaft were 3.642 m apart. A theodolite was set up slightly off the line *AB* and a distance 6.165 m from the wire *B*. The angle *ACB* was found to be 121″. Calculate the rectangular distance from *C* to the line produced.

(ICE Ans. 9.7 mm)

5 (a) From an underground traverse between 2 shaft wires *A* and *D*, the following partial coordinates in feet were obtained.

	ΔE (ft)	ΔN (ft)
AB	150.632	−327.958
BC	528.314	82.115
CD	26.075	428.862

Transform the above coordinates to give the grid coordinates of Station *B* given that the grid coordinates of *A* and *D* were as follows.

	E (m)	N (m)
A	520 163.462	432 182.684
D	520 378.827	432 238.359

(b) Compute the separate parts which make up the scale factor (K).

(TP Ans. 520 209.363 m E, 432 082.481 m N; conversion factor ft/m 0.3048; local scale factor 0.999 7780; survey error factor 1.002 1684)

2.11 The gyroscopic theodolite

This instrument is now universally used for the determination of true north. Whilst it does not produce the accuracy of astronomical surveying methods, for most purposes in engineering surveying it has been proved acceptable and is well established in such large organisations as the National Coal Board. The following two main types are available.

1) The suspended type, e.g. Wild GAK/1, MOM Gi series, Fennel series, where the gyroscope is attached above the transit axis of the theodolite.
2) The floating type, e.g. PIM, the precision indicator of the meridian, essentially an integrated instrument fixed co-axially below the theodolite.

The latter type is semi-automatic and as such not considered here.

2.11.1 Basic principles of the suspended gyroscope

The gyroscope is suspended by a metallic tape whilst the rotor itself is set in a gymbol, allowing freedom of movement in two planes but with the spin axis constrained due to the effect of gravity. The rotor has most of its mass concentrated at its rim and is driven at a speed of about 22 000 rpm with its spin axis horizontal (Fig. 2.26). If the spin axis is pointing due N/S at the time of 'spin up' it will remain in this position, but due to the general lack of precision in the alignment the earth's rotation will cause a slight inclination in the spin axis and a restoring gravity couple will be initiated. There is thus a resultant alignment of the spin axis which will oscillate about true north.

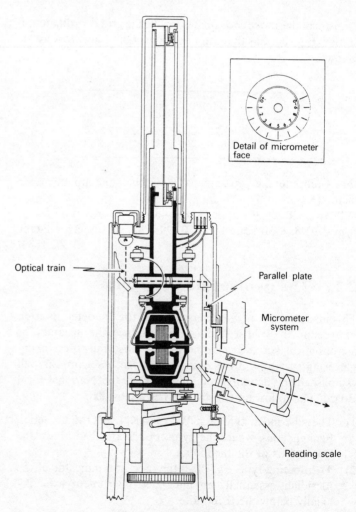

Fig. 2.26 RSM – Wild modified GAK/1 gyro attachment

This precession is very slow and although some form of damping may be present the ultimate time for it to come to rest would be considerable and as such the simple harmonic motion is used to effect various solutions.

The damped simple harmonic oscillation equation is given as

$$\varepsilon = A e^{-\mu t/2} \sin 2\pi t/T \qquad (2.41)$$

where ε = the angular displacement from the oscillation centre

μ (secs^{-1}) = the damping constant

t (secs) = the time measured from a given instant

T (secs) = the period of oscillation

The oscillation period is dependent on latitude (ψ)

$$T = T_0 \sec \psi \qquad (2.42)$$

where T_0 = the oscillation period between transits at the equator

T_r = the oscillation period between reversals

$$= T \sqrt{(1+c)} \qquad (2.43)$$

where $c = c_0 \sec \psi$

= the tape torque (K_1)/the precession torque (K)
$$\qquad (2.44)$$

$$K = K_0 \cos \psi \qquad (2.45)$$

These values are given for two gyroscopes at latitude 52° 57′ N

	T_0 (secs)	T	T_r	c_0	c
GAK/1	337.3	434.54	493.17	0.1737	0.2883
MOM GiC2	335.8	432.61	470.77	0.1110	0.1842

In equation 2.41 the damping term is $e^{-\mu t/2}$. If this term is linear, as in the GAK/1 and MOM GiC2, then

$$e^{-\mu t/2} = 1 - \mu t/2$$

If the damping term is parabolic,

$$e = 1 - (\mu t/2) + (\mu^2 t^2/8)$$

This factor may be found by noting the horizontal scale readings $U_1 U_2$ and U_3 at reversals $r_1 r_2$ and r_3 (see p. 101). Then

$$e^{-\mu T/4} = (U_3 - U_2)/(U_1 - U_2) \quad \text{where } T = 2t \qquad (2.46)$$

2.11.2 The zero position of the tape

The tape zero position is defined by Wild as the position of rest of the oscillating system relative to the instrument with the gyro not spinning. Their hand book suggests that a physical correction be made to the tape suspension screws. Thomas recommends that the tape zero be determined before and after 'spin up' of the gyro. The gyro is allowed to oscillate with the spinner at rest and is thus affected by the active torque of the tape. It is suggested that four reversals are recorded as amplitude readings on the gyro-mark scale and the centre of oscillation is given by Schuler's mean

$$\delta = (\delta_1 + 3\delta_2 + 3\delta_3 + \delta_4)/8 \qquad (2.47)$$

A correction factor is

$$c\,s\,\delta_m \qquad (2.48)$$

where $\delta_m = (\delta_{\text{before}} + 3\delta_{\text{after}})/4 \qquad (2.49)$

$s=$ the angular value of 1 division on the scale

2.12 Alternative observational methods

There are four alternative observational methods.
1) The reversal point system.
2) The transit system recommended by Wild.
3) The transit system recommended by the Royal School of Mines.
4) The amplitude system recommended by the Royal School of Mines.

Note that the last two techniques require modification to a GAK/1 incorporating a micrometer to read more precise values on the gyro-mark scale.

2.12.1 The reversal point method

The telescope is pre-orientated to within $\pm 2°$ of North (see Fig. 2.27). Horizontal scale readings (U_i) are taken at reversal points by smoothly tracking the gyro-mark. The centre of the oscillation is given by Schuler's mean as

$$U_1 = (u_1 + 2u_2 + u_3)/4 \qquad (2.50)$$
$$U_2 = (u_2 + 2u_3 + u_4)/4 \qquad (2.51)$$
then $\quad U = (u_1 + 3u_2 + 3u_3 + u_4)/8 \qquad (2.52)$

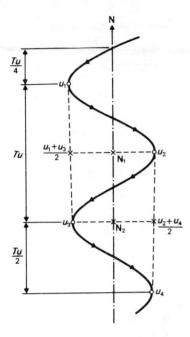

Fig. 2.27 Reversal point method (courtesy of Wild UK)

If the damping is linear this value represents the least square solution or, if parabolic, then the equation represents an exact solution. The relative position of the preceding values are seen in Fig. 2.28(a). To obtain the bearing of the base line (O/RO), in Fig. 2.28(b),

$$\phi_{RO} = \theta - U(+360) \tag{2.53}$$

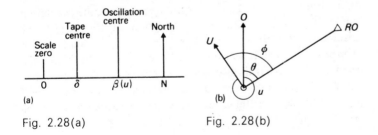

Fig. 2.28(a)   Fig. 2.28(b)

For example, given the following horizontal circle readings:

to RO $\theta = 54°\,20.6'$ (mean FL/FR)
to North $U = 357°\,01.1'$ (Schuler's mean)
    $\phi_{RO} = 54°\,20.6' + 360 - 357°\,01.1'$
        $= 57°\,19.5'$

Example of booking 2.16a *Reversal point method*
Theodolite No. T/16 (3)   Date. 1.2.1981

Gyro No. 1234   Observer. A.A.

Observation station base to Reference station X/42

RO   FL   01° 34.6' $c = 0.1783$
      FR   181° 34.7' $s = 2' 00''$
Mean (θ)   01° 34.65'
Reversals (u)   Tape zero (δ)

L	R	L	R
(u_1) 166° 35.7'	(u_2) 163° 26.9'	(δ_1) +6.7	(δ_2) −5.0
(u_3) 166° 35.2'	(u_4) 163° 27.1'	(δ_3) +6.0	(δ_4) −4.2

U' = 165° 01.13'   Schuler's mean $\delta = 0.688$
$-cs\delta$ = − 00.25'
U = 165° 00.88'

RO   Circle reading (θ)   01° 34.65'
      (if required)   + 360° 00.00
                          ─────────
                          361° 34.65'
Gyro-orientation (U) − 165° 00.88'
Calibration constant (E)   2.50'
                          ─────────
Geographical azimuth   196° 31.27'

ϕ_{RO} 196° 31' 16''

2.12.2 The transit method (A)

This method (recommended by Wild) depends upon the fact that within about ±15' of arc from the meridian, the oscillation curve is practically linear at the point of transit, i.e. the displacement of the gyro-mark is directly proportional to time. The horizontal circle reading is noted as N' and with the aid of a stop watch with a trailing hand (or preferably with a digital stop watch), the time (t_n) is recorded as the gyro-mark oscillates and transits the zero of the graduated scale. The adjustment to the assumed meridian value (N') is given as:

$$N = N' + \Delta N \qquad (2.54)$$

where

$$\Delta N = c_f a \Delta t \qquad (2.55)$$

c_f = the proportionality factor, dependent on latitude (ψ)

a = the amplitude

Δt = the difference in swing time

Note that in Fig. 2.29, if $a_e > a_w$ then N' is west of N, i.e. ΔN is +ve.

$$a = (a_e + a_w)/2 \quad (2.56)$$
$$\Delta t = (t_2 - t_1) - (t_3 - t_2) = T_E - T_W \quad (2.57)$$

If ω is a constant angular velocity over the central part of the curve and ΔN is the displacement of the theodolite pointing from the meridian, then

$$T_E - \Delta N/\omega = T_W + \Delta N/\omega$$
$$\Delta N = \omega(T_E - T_W)/2 = (\omega/2)\Delta t \quad (2.58)$$

As the swing time is constant for a given latitude (ψ), $\omega \propto a$ and

$$\tfrac{1}{2}\omega = c_f a \quad \text{(or } c_f = \omega/2a)$$
$$\Delta N = c_f a \Delta t \quad \text{(equation 2.52)}$$

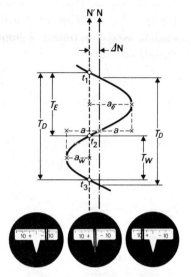

Fig. 2.29 Transit method. Positions of the gyro mark during transit method

The proportionality (compatibility) factor (c_f)
This value is constant for a given instrument at a given latitude (ψ) and is valid only for the linear range of the oscillation curve or about $\tfrac{1}{6}$ of the amplitude. It is recommended that Δt should not exceed 40 sec and ΔN should not exceed $0° 20'$ when $a = \pm 10$ units (i.e. Wild 100').
Theoretical values are given as follows.

$$c_f = s\pi(1 + c)/2T \quad \text{mins of arc/secs of time/div} \quad (2.59)$$
or $\quad c_f = s\pi T_r^2/2T^3 \quad \text{mins of arc/secs of time/div} \quad (2.60)$

For example, taking the GAK/1, $s = 10'$ and $T_r = 493.17$ secs. By equation 2.59,

$$c_f = 10\pi(1.2883)/(2 \times 434.5)$$
$$= 4.66 \times 10^{-2}$$

By equation 2.60,

$$c_f = 10\pi(493.17^2)/(2 \times 434.5^3)$$
$$= 4.66 \times 10^{-2}$$

Example of booking 2.16b   *Transit (Schwendener) method*   Date...1.2.81

Gyro No. 1235 $c_f = $ ...0.046... $s = $ 10'/div ....   Observation station...Base..

Observer...A.A. $c = 0.1783$ Reference station...X/42...

RO   FL   01° 34.6'   Tape zero (δ) +1.8 −1.4
     FR   181° 34.7'   +1.6 −1.3 $\delta = 0.138$

Mean (θ)   01° 34.65' $\underline{\delta U} = cs\delta = 0.1783 \times 10 \times 0.138 = \underline{0.25'}$

Circle reading N' $N = c_f a \Delta t$ $N = N' + \Delta N$		Time of transit m s	Swing time m s	Δt s	Amplitude a
N'	164° 58.90'	t_0 0 00.0 t_1 3 33.3 t_2 7 12.0 t_3 10 45.4 t_4 14 24.5	−3 33.3 +3 38.7 −3 33.4 +3 39.1	+5.4 +5.3 +5.7	+8.0 −7.7 +8.0 −7.8
ΔN	1.98		Mean +5.47		7.88
	165° 00.88'				
δU	− 00.25'				
U	165° 00.63'				

RO (θ)	01° 34.65'
(+360)	361° 34.65'
Gyro orientation (U)	165° 00.63'
	196° 34.02'
Calibration (E)	− 2.50'
Geographical N	196° 31.52'

$\phi_{RO} \ldots 196° 31' 31''.$

In practice the value of c_f may be found by making two separate measurements with the theodolite pointing in directions N'_1 and N'_2 each within 0° 15' East and West respectively. Then

$$N = N'_1 + c_f a_1 \Delta t_1 = N'_2 + c_f a_2 \Delta t_2$$
$$c_f = (N'_2 - N'_1)/(a_1 \Delta t_1 - a_2 \Delta t_2) \quad (2.61)$$

Observations for the determination of the compatibility factor (c_f) Following the booking process described previously the following mean values were recorded.

$$N'_1 = 164° 45.3' \quad \Delta t_1 = +33.4 \quad a_1 = 10.05$$
$$N'_2 = 165° 16.3' \quad \Delta t_2 = -31.7 \quad a_2 = 10.55$$

Then $c_f = (5°16.3' - 4°45.3')/[(33.4 \times 10.05)$
$$- (-31.7 \times 10.55)]$$
$$= 31.0'/670.11'' = 0.0463'/\text{sec/div}$$

Check N = 164° 45.3 + (0.0463 × 33.4 × 10.05) = 165° 00.8'
$$= 165° 16.3 + (0.0463 \times -31.7 \times 10.55)$$
$$= 165° 00.8'$$

2.12.3 The transit method (B)

This method (recommended by the Royal School of Mines—modified GAK/1) is similar to the previous method in that the times t_0, t_1, etc. are recorded as the gyro-mark passes through

(α_0), a selected graduation reading on the scale. At the reversal points α_1, α_2, etc. the amplitude values are read as in the previous method. Here the period of oscillation is derived as follows.

$$T = T_1 + T_2 \tag{2.62}$$

where $T_1 = \tfrac{1}{3}[(t_2 - t_0) + (t_3 - t_1) + (t_4 - t_2)]$ (2.63)

$$T_2 = \tfrac{1}{3}[(t'_2 - t'_0) + (t'_3 - t'_1) + (t'_4 - t'_2)] \tag{2.64}$$

Thomas then gives the oscillation centre

$$\beta_i = [\alpha_0 + \alpha_{ri}(p_i - 1)]/p_i \tag{2.65}$$
$$p_i = 2 \sin^2 X \tag{2.66}$$

where $\quad X = \dfrac{90}{T}(t_i - t_{i-1}) \tag{2.67}$

The values of T_1 and T_2 are taken with different α_0 positions (see Example 2.17), and mean values β and β' are derived.

2.12.4 The amplitude method

In this method (recommended by the Royal School of Mines—modified GAK/1), as with the previous method, the circle reading is recorded (pre-orientation $\pm 1°$ of North). The gyro-mark is allowed to oscillate about the gyro scale. (Note that if the oscillation exceeds the limits of the scale, it is damped until the amplitude falls within the scale values.) Four reversals are now observed on the scale as a_1, a_2, a_3 and a_4. The modification made to the GAK/1 (see Fig. 2.26) enables the values to be determined more precisely. A second set of reversals, recommended by Thomas, are a'_1, a'_2, a'_3 and a'_4 such that the circle reading is $u \pm 1'00''$. The mean values are obtained by the Schuler's equation,

$$\beta = (a_1 + 3a_2 + 3a_3 + a_4)/8$$

and $\quad \beta' = (a'_1 + 3a'_2 + 3a'_3 + a'_4)/8$

giving a mean value

$$N_m = U + s\omega \tag{2.68}$$

where

$$\omega = \beta(1 + c) - c\delta \tag{2.69}$$

$c = $ the instrument constant

$\delta = $ the tape zero $= (\delta_A + 3\delta_B)/4$

and

$$1 + c = (U' - U)/[s(\beta - \beta')] \tag{2.70}$$

(See Example 2.17.)

2.13 The instrument constant (E)

This represents the malalignment of the theodolite with:
a) the gyro-mark;
b) the north seeking vector.

The value E is given by

most probable bearing − observed bearing = $\phi_T - \phi_0$ (2.71)

i.e. $\phi_T = \phi_0 + E$

      = geographical azimuth of base line

2.14 Grid bearing of the base line

$$\phi_{\text{grid}} = \phi_T \pm \gamma \quad \text{(convergence)} \quad \text{(see p. 162)}$$

Example 2.17

Modified gyroscopic theodolite GAK/1

Transit/amplitude methods

Observation station Base to Reference station X/43   Date 18.2.81.

Formulae: Observer A.B.

$$s = 10 \text{ mins/div}$$

t_0, t_1 = consecutive transit times through α_0

α_0, α_r = approximate centre and reversal point readings

T = period of oscillation U = theodolite reading

$X° = 90(t_i - t_{i-1})/T$ $w = (\beta(1+c) - c\delta)$

$p_i = 2\sin^2 X$ N = circle reading at North

                                   $= U + sw$

β_i = oscillation centre

   $= (\alpha_0 + \alpha_{ri}(p_i - 1))/p_i$ β = mean value

$1 + c = (U' - U)/s(\beta - \beta')$

		Before	After		Before	After
RO	FL	356° 01′ 00″	356° 00′ 58″	Tape zero	7.10 −4.25	6.54 −3.80
	FR	176° 00′ 58″	176° 00′ 56″		7.04 −4.21	6.51 −3.75
	Mean (θ)	356° 00′ 58″			$\delta_B = 1.4075$	$\delta_A = 1.3650$
	Weighted mean = $(\delta_B + 3\delta_A)/4 = 1.3756$					

Transit observations (1)   Oscillation centre (β)

Transit times   Transit times

			α_0	2.00	p	β			α'_0	−1.00	p'	β'
t_0	0.0		α_1	13.73	0.991 886	1.9040	t'_0	0.0	α'_1	10.35	0.983 229	−1.1936
t_1	216.6		α_2	−9.95	1.009 200	1.8910	t'_1	215.4	α'_2	−12.74	1.014 972	−1.1732
t_2	435.6		α_3	13.72	0.992 607	1.9127	t'_2	435.2	α'_3	10.33	0.984 672	−1.1764
t_3	652.3		α_4	−9.94	1.007 036	1.9165	t'_3	650.8	α'_4	−12.73	1.014 972	−1.1728
t_4	871.0						t'_4	870.6				
$90/T = 0.206\,683$					Mean $\beta = 1.9061$						Mean $\beta' = -1.1790$	

Amplitude observations (2)
Reversal points $\beta = (a_1 + 3a_2 + 3a_3 + a_4)/8$

$$\begin{array}{cc} 13.74 & -9.93 \\ 13.73 & -9.92 \end{array} \quad \beta = 1.9025 \qquad \begin{array}{cc} 10.35 & -12.73 \\ 10.34 & -12.72 \end{array} \quad \beta' = -1.1925$$

(1) Observed value $c = 0.2966$
(2) Observed value $c = 0.2924$ Mean value 0.2945

	Amplitude		Transit	
Circle readings (U)	359° 45′ 00″	00° 25′ 00″	359° 45′ 00″	00° 25′ 00″
Correction (sw)	0° 20′ 34″	− 19′ 26″	20′ 38″	− 19′ 22″
Circle reading (N)	0° 05′ 34″	0° 05′ 34″	0° 05′ 38″	0° 05′ 38″
Bearing to RO ($RO - N$)	355° 55′ 24″	355° 55′ 24″	355° 55′ 20″	355° 55′ 20″
E factor	05′ 47″	05′ 47″	05′ 47″	05′ 47″
True bearing RO	356° 01′ 11″	356° 01′ 11″	356° 01′ 07″	356° 01′ 07″

356° 01′ 09″ Mean
1° 22′ 59″ Convergence
357° 24′ 08″ Grid bearing

Exercises 2.5

1 In determining the approximate direction of the meridian, the gyro-mark was tracked and the following theodolite readings were recorded at opposite reversal points.

RO (mean) 114° 22′
u_1 252° 15′
u_2 267° 42′

Calculate the approximate theodolite scale reading for North and the approximate bearing of the RO.

(Ans. 259° 58′, 214° 23′)

2 The following data was recorded in the transit method for the determination of the Grid bearing of a reference line AB.

Mean RO 144° 30.6′
Approximate orientation 267° 00.0′

Transit and amplitude values:

m	s	a
0	00.0	+15.0
3	40.7	−15.9
7	12.3	+15.0
10	53.4	−15.9
14	25.2	

The compatibility factor is given as 0.046′/sec/div, the instrument constant $E = -2.7$; and the convergence of meridians is $+1° 15.5'$. Calculate the Grid bearing of the line AB.
(1 div = 10′) (Ans. 238° 50.0′)

3 As part of reversal point observations, the following circle and scale readings were made:

Reversal readings	u_1	u_2	u_3	u_4
During spin up	142° 22′ 18″	139° 13′ 05″	142° 22′ 11″	139° 13′ 22″
Zero tape recordings	δ_1	δ_2	δ_3	δ_4
	26.1	−14.3	25.4	−13.6

Given that the scale readings for the observation of the RO was 34° 18′ 16″ and the instrument constant (E) was 3′ 14″, calculate the true bearing of the RO (assume that $c = 0.1783$, $s = 120″$).
(Ans. 253° 35′ 52″)

4 The gyro-theodolite was set up on a base line and the mean circle reading of this line was recorded as 84° 30′ 50″. Using the amplitude method, the following reversal points were recorded on the scale.

	δ_1	δ_2	δ_3	δ_4
Pre-spin up	10.9	−10.1	10.4	−9.4
	a_1	a_2	a_3	a_4
During spin up	28.1	−44.3	27.5	−44.0

Given that the observations were made at 52° 30′ N, $c_0 = 0.1110$, the value of 1 division on the scale was 120″ and the instrument factor (E) was +03′ 20″, calculate the true bearing of the reference line if the circle reading at the time of observation was 100° 30′ 26″. (Ans. 344° 23′ 26″)

5 As part of a major correlation survey an underground baseline was established and connected to the National Grid via a wire in one of the main shafts. A gyro-theodolite was set on this line and using the Reversal Point method the following data was recorded.

Horizontal circle readings	Remarks
016° 12′ 16″	RO on the base line
217° 50′ 10″	Left reversal
225° 32′ 20″	Right reversal
218° 00′ 44″	Left reversal
225° 09′ 50″	Right reversal

Subsequent computations provided the following additional information.

Convergence of the meridians	0° 20′ 30″
$(t-T)$ correction	0° 00′ 05″
Easting of the baseline	520 000 m

If the gyro-theodolite set up on the surface baseline of known azimuth 176° 32′ 42″ produced a value of 176° 12′ 22″, determine the National Grid bearing of the underground base line. (RICS/M Ans. 154° 29′ 37″)

6 Critically discuss the relative merits of the following correlation methods available to the minerals surveyor. (a) Using plumb wires in vertical shafts. (b) Using the gyro-theodolite.

7 In checking the correlation of two adjacent collieries, a gyro-theodolite set at A was used to determine the true bearing between station A at one colliery and station B at the other as 083° 53′ 50″. By previous observations, the Grid coordinates of these stations are as follows.

	E	N
A	366 180.25	428 156.62
B	369 252.75	428 460.52

If the approximate latitude of A was 53° 45′ discuss the validity of the correlation. (Assume the radius of the earth as 6370 km.)
(RICS/M)

8 (a) In the transit method of gyroscopic theodolite observations, ΔN, the adjustment to the circle reading N' is given as $\Delta N = ca\Delta t$ where $c =$ the proportionality factor, $a =$ the amplitude, $\Delta t =$ the time difference. Discuss, with the aid of a diagram, the implications of this formula.
(b) The following data was recorded in the transit method in order to determine the Grid bearing of a survey line.

$N'_1 = 0° 10.3′ \qquad a_1 = 10.42 \qquad \Delta t_1 = -32.2″$

$N'_2 = 359° 38.6′ \qquad a_2 = 10.21 \qquad \Delta t_2 = 34.6″$

$RO \quad FL \quad 44° 25.5′ \qquad FR \quad 224° 25.7′$

Meridian convergence $= +30.5′$

E factor $ = -3.5′$

Calculate the Grid bearing of the survey line.
(RICS/M Ans. 44° 57.7′)

Bibliography

ASHKENAZI, V., COUTIE, M. G., and SNELL, C., *Matrix Methods in Civil Engineering Networks*. Prepared for a Nottingham University Seminar (1968)

HODGES, D. J., and BROWN, J., *Underground and Surface Orientation Measurements with Gyrotheodolite Attachments*. The Institution of Mining Engineers (1972)

LILLEY, J. E., Least squares adjustment of dissimilar quantities. *Empire Survey Review*, **XVI**, 121 (1961)

THOMAS, T. L., The suspended gyroscope, Part 2, Theory of the suspended gyroscope. *Chartered Surveyor, Land, Hydrographic and Minerals Quarterly*, **2**, 3 (Spring 1975)

THOMAS, T. L., and ASQUITH, D., The suspended gyroscope, Part 1, Methods of finding north. *Chartered Surveyor, Land, Hydrographic and Minerals Quarterly*, **2**, 3 (Spring 1975)

THORNTON-SMITH, G. J., Least squares adjustment of dissimilar quantities. *Survey Review*, **XVIII**, 136 (1965)

SHEPHERD, F. A., *Engineering Surveying: Problems and Solutions*. Edward Arnold (1977)

SMITH, R. C. H., A modified GAK/1 gyro attachment. *Survey Review*, **XXIV**, 183 (1977)

3
Control measurements

The following three major forms of control measurement are considered here.

1) Optical distance measurement (ODM).
2) Electromagnetic distance measurement (EDM).
3) Trigonometrical heighting for vertical control.

3.1 Optical distance measurement

For relatively precise methods (i.e. 1/10 000 plus) there are two possible alternatives:

a) optical wedge tacheometry;
b) horizontal subtense bar tacheometry.

The former is now becoming less important with the advance in EDM; the manufacturers are replacing the optical wedge instruments with short range digital read-out EDM equipment attached to the theodolite or as 'total station' integrated instruments. The second method is still used, having recently become more popular again in the engineering field over distances less than 40 m. To achieve the required accuracy a 1″ theodolite is essential.

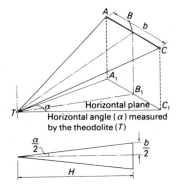

Fig. 3.1 Horizontal subtense bar system

3.2 Horizontal subtense bar tacheometry

The horizontal bar of known length (b), usually 2 m, is set normal to the line of sight TB (see Fig. 3.1). Targets A and C are successively sighted and the subtended angle (α) measured in the horizontal plane is recorded. The horizontal length is given by

$$TB_1 = H = (b/2) \cot(\alpha/2) \qquad (3.1)$$

If the bar is 2 m long then

$$H = \cot(\alpha/2) \quad \text{metres} \qquad (3.2)$$

The horizontal angle is measured several times (say 10) and, as the bar is horizontal, then the angle need only be measured on one face.

3.3 Factors affecting accuracy

3.3.1 The effect of an error in the subtended angle (α)

As
$$H = (b/2)\cot(\alpha/2) = b/[2\tan(\alpha/2)]$$
if the angle (α) is small then
$$(\alpha/2)_{rad} \simeq \tan(\alpha/2)$$
and $\quad H \simeq b/\alpha = 206\,265\, b/\alpha'' \quad$ (3.3)

This approximation should not be used to compute the length but has considerable advantages in the simplification of the analysis of the effects of the errors.

Partially differentiating the equations and writing in terms of the errors gives, using

$$H = (b/2)\cot(\alpha/2)$$
$$\partial H/\partial \alpha \equiv \delta H_\alpha/\delta \alpha$$
$$\delta H_\alpha = -(b/2)\cosec^2(\alpha/2)\delta\alpha/2 \quad (3.4)$$

and using
$$H = b/\alpha$$
$$\delta H_\alpha = -b\delta\alpha/\alpha^2 = -H^2 \delta\alpha/b \quad (3.5)$$

The error ratio can be given as
$$\delta H/H = \delta\alpha/\alpha \quad (3.6)$$

Example 3.1 The mean observed angle (α) to a 2 m bar is $1°\,40'\,39.5'' \pm 0.5''$. Calculate the horizontal length, the standard error in the length and the error ratio.

Note that $b = 2$ m.
$$H = \cot(\alpha/2) = \cot 0.8388° = 68.300 \text{ m}$$

By equation 3.4,
$$\delta H_\alpha = -\cosec^2(\alpha/2)\delta\alpha''/(2 \times 206\,265)$$
$$= \pm 0.5''/(2 \times 206\,265 \sin^2 0.8388°)$$
$$= \pm 0.0057 \text{ m} \quad (0.006 \text{ m})$$

By equation 3.5,
$$\delta H_\alpha = -H^2 \delta\alpha/b = (-68.300^2 \times 0.5'')/(206\,265 \times 2)$$
$$= \pm 0.0057 \text{ m} \simeq 0.006 \text{ m}$$
$$\delta H/H = \delta\alpha/\alpha = 0.0057/68.300 = 0.5/6039.5 \simeq 1/12\,000$$

Note that when $\alpha = 1.677\,639°$, $\alpha/2 = 0.838\,819°$

Then $\delta\alpha = 0.5'' = 0.000\,139°$ and $\delta\alpha/2 = 0.000\,070°$

and $H = \cot 0.838\,819° = 68.300\,415\,\text{m}$ (68.300 m)

$H + \delta H = \cot(\alpha + \delta\alpha)/2 = 68.294\,715\,\text{m}$ (68.295 m)

$\delta H = 0.005\,700\,\text{m}$ (0.006 m)

It can be seen from the above that the use of equation 3.5 gives the same numerical values but the analysis is simpler. To obtain the distance and the standard error both α and $\delta\alpha$ are halved. By equation 3.5,

$$\delta H_\alpha = -H^2 \delta\alpha/b$$

When $b = 2\,\text{m}$ and $\delta\alpha = 1''$, then

$$\delta H_\alpha = H^2/(2 \times 206\,265) \simeq H^2/400\,000 \qquad (3.7)$$

Example 3.2 To what accuracy should the subtense angle (α) be measured to a 2 m bar if the length of sight is approximately 50 m and the fractional error (error ratio) of 1/10 000 must not be exceeded?

By equation 3.6,

$$\delta H/H = \delta\alpha/\alpha = 1/10\,000$$
$$\delta\alpha = \alpha/10\,000 = b/(10\,000 \times H)$$
$$= (2 \times 206\,265)/(10\,000 \times 50)$$
$$= \pm 0.83''$$

Note that with glass arc 1" theodolites, the standard error of a single measurement of α should not exceed $\pm 1.5''$ and thus nine observations should give at least $\pm 0.5''$ and the standard error in the distance is dependent on $\pm 0.5/2$.

Example 3.3 If the measured angle (α) is approximately 2°, how accurately must it be recorded if the error ratio must not exceed 1/10 000?

$$\delta H/H = \delta\alpha/\alpha = 1/10\,000$$
$$\delta\alpha'' = 2 \times 3600/10\,000 = \pm 0.72''$$

Example 3.4 Assuming the angle (α) can be measured to $\pm 1''$ and the bar is 2 m, what is the maximum length that can be measured to an accuracy of 1/10 000?

$$\delta\alpha/\alpha = 1/10\,000$$

Then $\alpha = 10\,000'' = 2.7778°$ and

$$H = 2 \times 206\,265/10\,000 = 41.25\,\text{m}$$

3.3.2 The effect of an error in the length of the bar

Given $H = b/\alpha$ and partially differentiating and writing in terms of the errors gives

$$\delta H_b = \delta b/\alpha = H\delta b/b \qquad (3.8)$$

or $\quad \delta H_b/H = \delta b/b$

If this error ratio is not to exceed 1/10 000, then the bar must be 2.0000 m $\pm$ 0.2 mm. The bar is usually made of invar steel and is guaranteed by the manufacturers to 0.05 mm. It thus requires a change of 20° C to affect the bar by these limits and so for most practical purposes the bar is considered constant.

3.3.3 The effect of an error in the orientation of the bar

Let the bar be rotated as shown in Fig. 3.2. Consider half the bar AB rotated through an angle θ to $A_1 B$. The line of sight will thus be assumed to be at A_2. Then

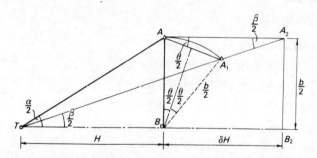

Fig. 3.2 Orientation of the bar

$$AA_1 = 2\,(b/2)\sin(\theta/2) = b\sin(\theta/2)$$
$$AA_2 = \delta H = AA_1 \sin(\theta/2 + \beta/2)/\sin(\beta/2)$$
$$= \frac{b\sin(\theta/2)\,[\sin(\theta/2)\cos(\beta/2) + \cos(\theta/2)\sin(\beta/2)]}{\sin(\beta/2)}$$
$$= b\sin^2(\theta/2)\cot(\beta/2) + b\sin(\theta/2)\cos(\theta/2)$$

But $\quad 2H \simeq b\cot(\beta/2)$

Then $\delta H = 2H\sin^2(\theta/2) + (b/2)\sin\theta$

Neglecting the second term, as $(b/2)\sin\theta$ is small, then

$$\delta H/H = 2\sin^2(\theta/2) \qquad (3.9)$$

and if the relative accuracy is limited to 1/10 000, then

$$\delta H/H = 1/10\,000 = 2\sin^2(\theta/2)$$

and $\qquad\qquad\qquad \theta = 0°\,48.6'$

As the bar is usually fitted with a sighting device, it can be orientated far more accurately than the above limit, and this non-rigorous analysis shows that this effect can be ignored.

3.3.4 Summary of the effects of the errors

It can be seen from the above that the whole system is almost entirely dependent upon the angle α. For a single bay measurement, with a standard error of $\pm 1''$ in the angle, the distance is limited to 41.25 m, if the error ratio is limited to 1/10 000. For many practical purposes, this distance may be increased to 75 m with a resultant accuracy of ± 10 mm, i.e. 1/7500.

To increase the range of the instrument, various processes may be used, and these are described in Section 3.4.

3.4 Methods used in the field

3.4.1 Serial measurement

With reference to Fig. 3.3,

$$T_1 T_2 = H_1 + H_2 + \ldots + H_n = H$$
$$= (b/2)[\cot(\alpha_1/2) + \cot(\alpha_2/2)$$
$$+ \ldots + \cot(\alpha_n/2)] \quad (3.10)$$

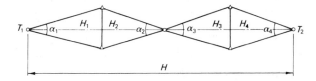

Fig. 3.3 Serial measurement

By equation 3.5,

$$\delta H_1 = -H_1^2 \delta\alpha_1 / b$$

Then the rms error is given by

$$\delta H = \sqrt{\Sigma \delta H^2}$$
$$= [(H_1^2 \delta\alpha_1/b)^2 + (H_2^2 \delta\alpha_2/b)^2 + \ldots]^{1/2} \quad (3.11)$$

If $H_1 = H_2 = H_n = H/n$ and $\alpha_1 = \alpha_2 = \alpha_n$, with $\delta\alpha_1 = \delta\alpha_2 = \delta\alpha_n$, then

$$\Sigma \delta H = \pm H^2 \delta\alpha / n^{3/2} b \quad (3.12)$$

If $b = 2$ m, $\delta\alpha = \pm 1''$ and $H = nH_1$, then

$$\delta H \simeq \pm H^2 / (400\,000\, n^{3/2}) \quad (3.13)$$
and $\quad \delta H/H \simeq H/(400\,000\, n^{3/2}) \quad (3.14)$

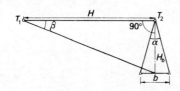

Fig. 3.4 Auxiliary base measurement

When $n = 2$ and $\delta H/H = 1/10\,000$ then

$$H = 400\,000 \times 2^{3/2}/10\,000 = 113\,\text{m}$$

3.4.2 Auxiliary base measurement

For lines in excess of 150 m, an auxiliary base may be set out at 90° to the main line (see Fig. 3.4). Angles α and β are measured.

$$H = H_b \cot \beta = (b/2) \cot (\alpha/2) \cot \beta \qquad (3.15)$$

If α and β are both small then applying the same error analysis,

$$\delta H = [(b\delta\alpha/\alpha^2 \beta)^2 + (b\delta\beta/\alpha\beta^2)^2]^{1/2} \qquad (3.16)$$

but as $H = b/(\alpha\beta)$

then $\quad \delta H = H[(\delta\alpha/\alpha)^2 + (\delta\beta/\beta)^2]^{1/2} \qquad (3.17)$

If $\alpha = \beta$ and $\delta\alpha = \delta\beta$ then

$$\delta H = \sqrt{2}\, H^{3/2}\, \delta\alpha/\sqrt{b} \qquad (3.18)$$

and $\quad H_b = bH/H_b = \sqrt{(bH)} \qquad (3.19)$

If $b = 2$ m and $\delta\alpha = \pm 1''$ then

$$\delta H = H^{3/2}/206\,265 \qquad (3.20)$$

and if $\delta H/H = 1/10\,000 = \sqrt{H}/206\,265$ then

$$H = 20.6265^2 \simeq 425\,\text{m}$$

and the sub-base

$$H_b = \sqrt{(2H)} = 29.2\,\text{m}$$

3.4.3 Central auxiliary base

For lines in excess of 400 m, a double bay system may be adopted with the auxiliary base in the middle (see Fig. 3.5). Then

$$T_1 T_2 = H = H_1 + H_2$$
$$= (b/2) \cot (\alpha/2) (\cot \beta_1 + \cot \beta_2) \qquad (3.21)$$

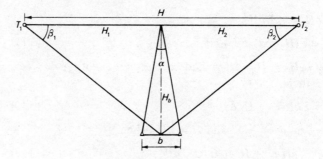

Fig. 3.5 Central auxiliary base

If α, β_1 and β_2 are all small then

$$H \simeq (b/\alpha)(1/\beta_1 + 1/\beta_2) \tag{3.22}$$

If $H_1 = H_2$ then $\beta_1 = \beta_2$, and if $\alpha = \beta$ and $\delta\alpha = \delta\beta_1 = \delta\beta_2$ then

$$\delta H = (\sqrt{3}H^{3/2}\delta\alpha)/(2\sqrt{b}) \tag{3.23}$$

If $b = 2\,\text{m}$ and $\delta\alpha = \pm 1''$ then

$$H = H_b^2 \tag{3.24}$$

and $\delta H \simeq H^{3/2}/335\,000$ (3.25)

If $\delta H/H = 1/10\,000$ then

$$\delta H/H = 1/10\,000 = \sqrt{H}/335\,000$$

$$H = 33.5 \simeq 1122\,\text{m}$$

and $\quad H_b = \sqrt{H} = 33.5\,\text{m}$

3.4.4 Auxiliary base perpendicularly bisected by the traverse line

With reference to Fig. 3.6,

$$T_1T_2 = H = H_1 + H_2$$
$$= (b/4)\cot(\alpha/2)[\cot(\beta_1/2) + \cot(\beta_2/2)] \tag{3.26}$$

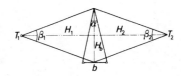

Fig. 3.6 Auxiliary base bisected by the traverse line

The analysis is similar to the above and the following equation is given, when $\alpha = \beta_1 = \beta_2$ and $\delta\alpha = \delta\beta_1 = \delta\beta_2$, as

$$\delta H = (\sqrt{3}H^{3/2}\delta\alpha)/(2\sqrt{b})$$

This is the same as equation 3.23 above.

3.4.5 With two auxiliary bases

With reference to Fig. 3.7,

$$H_b = (b/2)\cot(\alpha/2) \tag{3.27}$$
$$H_1 = H_b \cot\beta$$

and $\quad H = H_1 \cot\phi = (b/2)\cot(\alpha/2)\cot\phi\cot\beta \tag{3.28}$

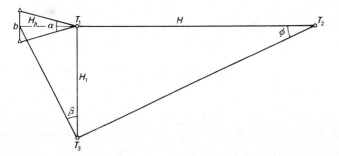

Fig. 3.7 Two auxiliary bases

The analysis as before gives, if $\alpha = \beta = \phi$ etc., then

$$\delta H = \sqrt{3} H^{4/3} \delta\alpha / b^{1/3} \qquad (3.29)$$

and if $b = 2$ m and $\delta\alpha = \pm 1''$ then

$$\delta H \simeq H^{4/3}/150\,000 \qquad (3.30)$$

and $\quad H_b = b^{2/3} H^{1/3} \qquad (3.31)$

If $\delta H/H = 1/10\,000$ then

$$H \simeq 15^3 = 3375 \text{ m} \quad \text{and} \quad H_b = 1.5866 \times 15 = 23.8 \text{ m}$$

Exercises 3.1

1 (a) Discuss the sources of error in horizontal subtense bar tacheometry showing the effect each will have on the final horizontal length.
(b) If the bar is 2 m long and the error of $\pm \delta\alpha$ is made in the subtense angle α, show that the error ratio $\delta H/H = H\delta\alpha/2$.
(c) If the error $\delta\alpha = \pm 1''$, what will be the error ratio $\delta H/H$ for a double bay measurement where the bar is placed midway between the theodolite stations which are 110.00 m apart?

(RICS/M Ans. 1/10 600)

2 (i) What do you understand by systematic and accidental errors in linear measurement, and how do they affect the assessment of the probable error? Does the error in the measurement of a particular distance vary in proportion to the distance or to the square root of the distance?
(ii) Assume you have a subtense bar the length of which is known to be exactly 2 metres and a theodolite with which horizontal angles can be measured to within a second of arc. In measuring a length of 600 m, what error in distance would you expect from an angular error of 1 second?
(iii) With the same equipment, how would you measure the distance of 600 m in order to achieve an accuracy of 1/5000?

(ICE Ans. ± 0.309 m)

3 (a) Describe, with the aid of sketches, the principles of subtense bar tacheometry.
(b) The sketch shows two adjacent lines of a traverse AB and BD with a common sub-base BC. Calculate the lengths of the traverse lines from the following data
Angles $\quad BAC = 5°\,10'\,30'' \quad CBA = 68°\,56'\,10''$

$YBX = 1°\,56'\,00'' \quad CDB = 12°\,54'\,20'' \quad DBC = 73°\,18'\,40''$
Length of bar 2 metres.

(EMEU Ans. AB, 631.96 m, BD, 264.78 m)

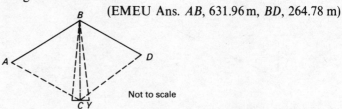

Not to scale

4 Describe the 'single bay' and 'double bay' methods of measuring linear distance by use of the subtense bar. Show that for a subtense bar

$$L = S/2 \cot \phi/2$$

where L = the horizontal length between stations; S = length of the subtense bar; ϕ = angle subtended by targets at the theodolite. Thereafter show that if $S = 2$ metres and the error of $\pm \Delta\phi$ is made in the measurement of the subtended angle, then $\Delta L/L = \pm \Delta\phi L/2$ where ΔL is the corresponding error in the computed length.

Assuming an error of $\pm 1''$ ($\Delta\phi$) in the measurement of the subtended angle what will be the fractional error at the following lengths? (a) 50 metres (b) 100 metres (c) 500 metres.
(LU Ans. (a) 1/8250 (b) 1/4125 (c) 1/825)

5 A colliery base line AB is unavoidably situated on ground where there are numerous obstructions which prevent direct measurement. It was decided to determine the length of AB by the method illustrated in the figure where DE is a 50 metre band hung in catenary with light targets attached at the zero and 50 metre marks. From the approximate angular values shown, determine the maximum allowable error in the measurements of the angles such that the projection of error due to these measurements does not exceed (a) 1/200 000 of the actual length CD when computing CD from the length DE and the angle DCE, and (b) 1/100 000 of the actual length AB when computing AB from the angles $ACD, CDA, BDC,$ and DCB and the length DC. For this calculation, assume that the length DC is free from error. (RICS/M (a) 0.3'' (b) 0.9'')

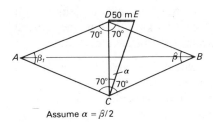

3.5 Electromagnetic distance measurement (EDM)

With the improvement in EDM many of the basic surveying techniques have changed, particularly as the linear measurements are now as good as, if not better than, the angular measurements.

3.5.1 Principles of EDM measurement

An EDM instrument generates an electromagnetic wave, within or near to the visible spectrum, with a modulated wave superimposed upon it for phase comparison purposes (see Fig. 3.8).

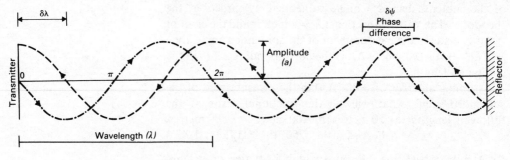

Fig. 3.8

The wavelengths of the carrier waves are given approximately as follows.

a) microwaves $\quad \lambda \simeq 1 \times 10^{-3}$ m to 1×10^{-1} m
b) visible light $\quad \lambda \simeq 3$ to 7×10^{-7} m (i.e. blue to red)
c) infrared $\quad \lambda \simeq 8 \times 10^{-7}$ m to 1×10^{-6} m
d) laser (He–Ne) $\lambda \simeq 6 \times 10^{-7}$ m

Manufacturers adopt a measuring frequency (f), a carrier wavelength (λ) and a basic measuring unit for the modulated wave (u). With the exception of the Mekometer ME 3000, the wavelength is not constant in operation but related to the frequency generated. In vacuo, the velocity of the propagation of electromagnetic waves is constant for all wavelengths and frequencies; i.e. $v_0 = 299\,792.5 \pm 0.4\,\text{km s}^{-1}$, but the general equation for velocity of propagation in the atmosphere is

$v = f\lambda = v_0/n'\quad$ the velocity in metres per second
or $\quad \lambda = v_0/(n'f) \quad$ the wavelength in metres
where $n' = v_0/v$, the refractive index, which in air is dependent upon the gaseous components: air temperature, pressure and humidity.

Cauchy's equation is given as

$$n = A + B/\lambda^2 + C/\lambda^4 \tag{3.32}$$

where A, B and C are constants for given atmospheric conditions. For dry air at 0° C, 760 mm Hg (1013.25 m bars) pressure and with 0.03 % CO_2, the refractive index becomes

$$n_s = 1 + (287.604 + 1.6288/\lambda^2 + 0.0136/\lambda^4)10^{-6} \tag{3.33}$$

where λ is given in μm, i.e. 1×10^{-6} m.

The measuring process is dependent upon phase comparison

3.5 ELECTROMAGNETIC DISTANCE MEASUREMENT (EDM)

between a travelling and standing wave and, as this comparison cannot be resolved better than $1/1000$ of λ, then a modulated wave (which is not sinusoidal) is superimposed upon the carrier wave. The measuring wavelength, or unit length, is chosen as a convenient number dependent upon the frequency generated. To take account of this, a group refractive index (n_g) is used as

$$n_g = A + 3B/\lambda^2 + 5C/\lambda^4 \qquad (3.34)$$

and thus equation 3.33 becomes

$$n_{gs} = 1 + (287.604 + 4.8864/\lambda^2 + 0.0680/\lambda^4)10^{-6} \quad (3.35)$$

This is the equation used in all subsequent calculations, and modified as

$$N_{gs} = (n_{gs} - 1)10^6 = 287.604 + 4.8864/\lambda^2 + 0.068/\lambda^4$$
$$(3.35a)$$

The refractive index under various atmospheric conditions is given by Barrell and Sears as

$$n_{gt} = 1 + \frac{273.2(n_{gs}-1)p}{760(273.2+t)} - \frac{15.02 E \times 10^{-6}}{273.2+t} \qquad (3.36)$$

$$= 1 + 0.359\,474\,(n_{gs}-1)p/T - 15.02E \times 10^{-6}/T$$
$$(3.36a)$$

where p = the atmospheric pressure (mm Hg)
$T = 273 + t$ (K)
E = the water vapour pressure
(mm Hg) $= e' - (c/755)p(t - t_w)$
e' = the saturated water vapour pressure at t_w (derived from tables)
t_w and t are the wet and dry temperature (°C)
c is a constant given for a wet bulb covered with water as 0.50 and covered with ice as 0.43.

Example 3.5 Given the standard refractive index of visible light in air at $0\,°C$ and 760 mm Hg (CO_2 content 0.03%) as 1.000 3045, find the refractive index at 25 °C and 730 mm Hg. Saturated vapour pressure at 25 °C is 23.7 mm Hg.

By equation 3.36a,

$$n_{gt} = 1 + 0.359\,474\,(n_{gs}-1)p/T - 15.02 E \times 10^{-6}/T$$

The second term is given by

$$0.359\,474 \times 3.045 \times 10^{-4} \times 730/298 = 2.681 \times 10^{-4}$$

The third term is given by

$$-15.02 \times 23.7 \times 10^{-6}/298 \qquad = -0.012 \times 10^{-4}$$

Then

$$n_{gt} = 1 + 0.000\,2681 - 0.000\,0012 \quad = 1.000\,2669$$

It can be seen from this that the effects of the third term, involving the water vapour pressure, is very small and in most cases can be neglected.

The effect on (n) of changes in following the conditions.

		Approximate changes (ppm) Visible light	Microwaves
Temperature	+1°C	−1.00	−1.25
Atmospheric pressure	+1 mm Hg	+0.40	+0.40
Water vapour pressure	+1 mm Hg	−0.05	+6.60

For standardisation purposes, manufacturers choose standard conditions and thus a standard refractive index (n_a) upon which their measurements are based.

By equation 3.36a,

$$N_a = (n_a - 1)10^6$$
$$= N_{gs} \times 0.359\,474 p_a/T_a - 15.02 E \times 10^{-6}/T_a$$
$$= N p_a/T_a - 15.02 E \times 10^{-6}/T_a \quad (3.37)$$

where $N = N_{gs} \times 0.359\,474$.

AGA quote standard values of N for various wavelengths λ:

λ (m)		N
0.5500	Mercury vapour lamp	109.460
0.5650	Standard lamp	109.129
0.6328	Red laser light	107.925
0.9200	Infrared	105.496
0.9300	Infrared	105.450

Example 3.6 Find the standard group refractive index for a carrier wave of white light where $\lambda = 0.5650$ μm.

By equation 3.35a,

$$N_{gs} = 287.604 + 4.8864/0.565^2 + 0.068/0.565^4$$
$$= 303.58 \quad \text{and} \quad n_{gs} = 1.000\,303\,58$$
$$N = 303.58 \times 0.359\,474 = 109.129 \quad \text{(see AGA table)}$$

'Correction' to the measured length
If, during the measurement, the atmospheric conditions are not standard then a 'correction' must be made to the recorded value (D). The reduced value is then

$$D' = D\,n_a/n_{gt} \quad (3.38)$$

where n_a is the standard refractive index for the instrument, and

n_{gt} is the refractive index at the time of measurement. The 'correction' is then

$$c = D' - D = D(n_a/n_{gt} - 1)\,\text{m} \tag{3.39}$$
$$= D(N_a - N_{gt})10^{-6} \tag{3.39a}$$

The correction factor $(N_a - N_{gt})$ given in parts per million (ppm) is frequently derived as a scaled value from a nomograph or from a manufacturer's slide rule. With many modern instruments, this value is dialled into the instrument and the distance is automatically adjusted and displayed as the reduced value. The measurement of the atmospheric conditions is difficult in practice and for more precise measurements (preferably as reciprocal observations), temperature, pressure and humidity should be measured at both ends of the line.

The Essen and Froome formula for refractive index of radio waves is given as

$$N_t = (n_t - 1)10^6 = \frac{103.49\,(p - E)}{T} + \frac{86.26\,E(1 + 5748/T)}{T} \tag{3.40}$$

As the radio wave type instrument is mainly confined to long range observations no further consideration is given here.

Example 3.7 An AGA Geodimeter Model 6 is standardised at $-4°\text{C}$, 760 mm Hg, with an effective wavelength of $0.5500\,\mu\text{m}$. A distance of 2535.365 m is recorded at a temperature of $14°\text{C}$, and at an atmospheric pressure of 740 mm Hg. Calculate the correction due to the atmospheric conditions and thus the reduced length (see Fig. 3.8).

By equation 3.35a,

$$N_{gs} = 287.604 + 4.8864/0.55^2 + 0.068/0.55^4 = 304.51$$

By equation 3.36a,

$$N_{gt} = 304.51 \times 0.359\,474 \times 740/287 = 282.24$$

By equation 3.37,

$$N_a = 304.51 \times 0.359\,474 \times 760/269 = 309.26$$

The 'correction' is then

$$c \simeq D(N_a - N_{gt}) = 2535.365\,(309.26 - 282.24) \times 10^{-6}$$
$$= 2535.365 \times 27.02\,\text{ppm} = 0.068\,\text{m}$$

and the reduced length is $= 2535.365 + 0.068 = 2535.433\,\text{m}$.

In most short and medium distance instruments the wave is returned back to the transmitter by using retro-directive reflectors (frequently the corner of a glass cube). If the reflector is moved through a distance X, then the wave will travel through a distance $2X$. The effective wavelength is then of length

$u = \lambda/2$. By measuring electronically or mechanically the phase shift $\delta\psi$ between the outgoing and incoming signals, a residual part of the wavelength $\delta\lambda$ may be recorded. The displacement of a single wave of amplitude a (metres) and frequency f (cycles/sec) at a time t (seconds) is given as

$$y = a \sin(2\pi ft) \tag{3.41}$$

and a second wave with a phase shift $\delta\psi$ is then

$$y = a \sin(2\pi ft + \delta\psi) \tag{3.42}$$

Note that the phase shift may be expressed as part of the wavelength $(\delta\lambda)$, i.e. as $\delta\psi \propto \delta\lambda$; then $\delta\psi = k\,\delta\lambda$. When the waves are in phase,

$$\delta\psi = 2\pi_{\text{rad}}$$

i.e. $\delta\lambda = \lambda$ and $k = \delta\psi/\delta\lambda = 2\pi/\lambda$.

Generally,

$$\delta\psi = 2\pi\delta\lambda/\lambda \tag{3.42a}$$
$$\text{and} \quad \delta\lambda = \lambda\,\delta\psi/2\pi \tag{3.42b}$$

This is the connecting equation between the phase and the path difference.

The resultant wave displacement, by superimposing one wave upon another, is given by

$$y_r = a[\sin(2\pi ft) + \sin(2\pi ft + \delta\psi)] \tag{3.43}$$
$$= 2a \cos(\delta\psi/2) \sin(2\pi ft + \delta\psi/2) \tag{3.43a}$$

where $2a \cos(\delta\psi/2)$ is the resultant amplitude.

When $\delta\psi = 2n\pi$, the resultant amplitude is a maximum $= +2a$ and the path difference is then $\delta\lambda = 2n\pi\lambda/2\pi = n\lambda$. Note that n is a positive integer.

When $\delta\psi = 2\pi(n+0.5)$, the resultant amplitude is a minimum, and the path difference is then

$$\delta\lambda = 2\pi(n+0.5)\lambda/2\pi = (n+0.5)\lambda$$

This path difference may be used to determine the distance as follows.

$$2D = \dot{n}\lambda + \delta\lambda' = n\lambda + \delta\psi\,\lambda/2\pi \tag{3.44}$$
$$\text{then} \quad D = n(\lambda/2) + \delta\lambda = nu + \delta u \tag{3.45}$$

where $u = \lambda/2$, i.e. half the modulated wavelength, and $\delta u = u\,\delta\psi/2\pi$.

As the value of n cannot be measured directly, it is necessary to introduce changes in frequency and thus changes in wavelength, followed by the solution of simultaneous equations. These principles are illustrated as follows by the analogous use of three tapes $(t_1, t_2 \text{ and } t_3)$ having lengths $(u_1, u_2 \text{ and } u_3$ respectively); see Fig. 3.9.

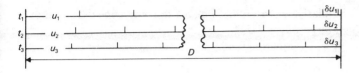

Fig. 3.9

With tape t_1 of length u_1, then

$$D = n_1 u_1 + \delta u_1$$

With tape t_3 of length u_3, then

$$D = n_3 u_3 + \delta u_3$$

Let $u_1 = 5.000$ m and $u_3 = 20u_1/21 = 4.761\,905$ m. Then, if $\delta u_1 < \delta u_3$, $n_1 = n_3 = n$, when $D \leqslant 100$ m, i.e. $20u_1 = 21u_3 = 100$ m and

$$D = nu_1 + \delta u_1 = nu_3 + \delta u_3$$

Here $n = (\delta u_3 - \delta u_1)/(u_1 - u_3) = 21(\delta u_3 - \delta u_1)/5$

$$= 4.2(\delta u_3 - \delta u_1) \quad (3.46)$$

and

$$D = 21(\delta u_3 - \delta u_1) + \delta u_1 \quad \text{or} \quad 20(\delta u_3 - \delta u_1) + \delta u_3 \quad (3.47)$$

Note that if $\delta u_1 > \delta u_3$, then $n_3 = n_1 + 1$.

Similarly, the second tape may now be introduced, i.e. t_2 of length $u_2 = 400u_1/401 = 4.987\,531$ m.

Provided $D \leqslant 2000$ m, then if $\delta u_1 < \delta u_2$, $n_1 = n_2 = n'$, i.e. $400u_1 = 401u_2 = 2000$ m. Now,

$$n' = (\delta u_2 - \delta u_1)/(u_1 - u_2) = 401(\delta u_2 - \delta u_1)/5$$

$$= 80.2(\delta u_2 - \delta u_1) \quad (3.48)$$

$$D = 401(\delta u_2 - \delta u_1) + \delta u_1 \quad \text{or} \quad 400(\delta u_2 - \delta u_1) + \delta u_2 \quad (3.49)$$

Note that if $\delta u_1 > \delta u_2$, then $n_2 = n_1 + 1$.

Example 3.8 Let $\delta u_1 = 2.263$ m and $\delta u_3 = 4.168$; then $\delta u_3 - \delta u_1 = 1.905$ m. By equation 3.48,

$$n = 4.2 \times 1.905 = 8.001 \simeq 8$$

$$D = (8 \times 5.000) + 2.263 = 42.263 \text{ m}$$

Also, $D = (8 \times 4.761\,905) + 4.168 = 42.263$ m

or, by equation 3.49,

$$D = (21 \times 1.905) + 2.263 = 42.268 \text{ m}$$

also $D = (20 \times 1.905) + 4.168 = 42.268$ m

The difference between these two sets of values is due to the limitation in the accuracy of the value of $(\delta u_3 - \delta u_1) = 1.905$ and thus it is necessary to quote this value to 5×10^{-5} m in order that the measurement can be quoted to 1 mm.

Similarly, let $\delta u_2 = 4.857$ m, made at the same time as above; then $\delta u_2 - \delta u_1 = 2.594$.

Now, by equation 3.48,

$$n = 80.2 \times 2.594 = 208.03 \simeq 208$$

Then $D = (208 \times 5.000) + 2.263 = 1042.263$ m
also $\quad D = (208 \times 4.987\,531) + 4.857 = 1042.263$ m

or by equation 3.49,

$$D = (401 \times 2.594) + 2.263 = 1042.457 \text{ m}$$

also $\quad D = (400 \times 2.594) + 4.857 = 1042.457$ m

As before, $\delta u_2 - \delta u_1 = 2.594$ is not sufficiently accurate and here the value would need to be quoted to an accuracy of 2.5×10^{-6} m to attain a measurement to 1 mm. To overcome these deficiencies and to simplify the computation in the field AGA recommends the following system.

Let $(\delta u_2 - \delta u_1) = (L_2 - L_1) = A$ and let $(\delta u_3 - \delta u_1) = (L_3 - L_1) = B$. Then let E be the largest multiple of 100 m contained in the distance and let F be the largest multiple of 5 m in the remaining part of the distance (i.e. less than 100 m). Then $F = 21B$ to the nearest 5 m and $E = 400A - 21B$ to the nearest 100 m. Thus

$$D' = (E + F) + L_1 \tag{3.50}$$

$$D'' = 400(E+F)/401 + L_2 = (E+F) + L_2 - (E+F)/401$$
$$D'' = (E+F) + L_2 - k_2(E+F) \tag{3.51}$$

where $k_2 = 0.002\,493\,766 = 1/401$.

$$D''' = E + 20F/21 + L_3 = (E+F) + L_3 - F/21$$
$$D''' = (E+F) + L_3 - K_3 F \tag{3.52}$$

where $k_3 = 0.047\,619 = 1/21$.

$$D = (D' + D'' + D''')/3 \tag{3.53}$$

In the example, $F = 21 \times 1.905 =$ (say) 40, $E = 400 \times 2.594 =$ (say) 1000. Then

$$D' = 1040 + 2.263 = 1042.263 \text{ m}$$
$$D'' = 1040 + 4.857 - 2.593 = 1042.264 \text{ m}$$
$$D''' = 1040 + 4.168 - 1.905 = 1042.263 \text{ m}$$
$$D = 1040 + (0.263 + 0.264 + 0.263)/3 = 1042.263 \text{ m}$$

Note that in the Model 6 Geodimeter, the distance is given as

$$D = nu + (R - C) + k \tag{3.54}$$

where $R =$ the reflector delay readout value

$C =$ the calibration delay readout value

$R - C = \delta u$, the residual length

$k =$ the zero error made up of the (geodimeter + reflector) constant.

3.5.2 Accuracy of EDM instruments

Manufacturers of this equipment specify the accuracies of their instruments as follows.

1) a mean square error (standard error) in the measurement e.g. ± 5 mm;
2) a proportional error due to variations in frequency and the effects of meteorological data; e.g. ± 2 parts per million of the distance D, i.e. ± 2 ppm or $2 \times 10^{-6} D$.

3.5.3 Calibration of the instruments

There are three sources of systematic errors which may be considered.

1) zero error;
2) cyclic error;
3) proportional error.

1) *Zero error*   This is the constant error (k) which is made up of two parts, the instrument error and the reflector error. It may be determined by the use of the instrument on a base line of known length. Therefore

$$k = \text{measured length } (d) - \text{known length } (l)$$
$$= d - l$$

Alternatively, a multi-pillared baseline may be used in which the instrument is set at various points along this straight line and measurements are recorded to the various reflector stations. Taking the minimum conditions, which involves three stations (Fig. 3.10), the observation equations are

$$l_1 + k = d_1$$
$$l_2 + k = d_2$$
$$(l_1 + l_2) + k = d_3$$

Then $\quad k = d_1 + d_2 - d_3$

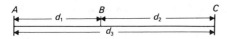

Fig. 3.10

Schwendener gives the general equation

$$C = \frac{D - [d]}{n - 1} = -k \qquad (3.55)$$

where C = the correction to the measured length
D = the overall measured length
$[d]$ = the sum of the measured bay lengths
n = the number of bays.

This provides a unique solution for k and for each of the bay lengths, in which the observational errors are attributed to k.

To provide a more accurate solution, a combination of lengths is required in which the application of a least square solution entails the formation of normal equations which, when solved, gives the most probable values of the lengths together with the 'zero' error k (see Example 3.9).

2) *Cyclic error* This is determined by the movement of the reflector by systematic increments over half a wavelength. It is seldom significant when the instruments are regularly serviced by the manufacturer.

3) *Proportional error* This is quoted, as above, as largely due to the systematic errors in the determination of the atmospheric refraction adjustment and some irregularity in the generated frequencies. It is difficult to detect and is generally overcome by comparison with well calibrated baselines when the reciprocal observations enable the atmospheric effect to be detected.

Example 3.9 An EDM instrument is thought to have a 'zero' error and to test this the following measurements were made on a multi-pillar base line. $AB = 29.972$ m, $AC = 80.473$ m, $AD = 210.820$ m, $BC = 50.475$ m, $BD = 180.830$ m, $CD = 130.330$ m. Calculate the most probable values of the distances between the pillars and the 'zero' error.

This example is used to illustrate the difference between the unique solution using the minimum amount of data and a least square solution using all the data.

(a) Using a minimum amount of data, i.e. AB, AC and BC.

Let $AB = a$, $BC = b$ and the zero error $= k$, then the observational equations are given as

$$d_1 = a \qquad + k = 29.972$$
$$d_2 = a + b + k = 80.473$$
$$d_3 = \qquad b + k = 50.475$$

Solving these equations gives the unique values as

$$a = 29.998 \text{ m}$$
$$b = 50.501 \text{ m}$$

and the 'zero error' $k = -0.026$ m.

(b) Using all the data, let the extra length $CD = c$; then the observational equations become

$$a \qquad\quad + k = \quad 29.972 + r_1$$
$$a + b \qquad + k = \quad 80.473 + r_2$$
$$a + b + c + k = 210.820 + r_3$$
$$\quad b \qquad\quad + k = \quad 50.475 + r_4$$

$$b + c + k = 180.830 + r_5$$
$$c + k = 130.330 + r_6$$

Applying the least square condition, the normal equations then become (see p. 19),

$$3a + 2b + c + 3k = 321.265$$
$$2a + 4b + 2c + 4k = 522.598$$
$$a + 2b + 3c + 3k = 521.980$$
$$3a + 4b + 3c + 6k = 682.900$$

Solving these equations simultaneously gives the most probable values as

$$a = 29.9937 \quad \text{say } 29.994 \text{ m}$$
$$b = 50.4985 \quad \text{say } 50.499 \text{ m}$$
$$c = 130.3513 \quad \text{say } 130.351 \text{ m}$$

and the 'zero' error $k = -0.0215$ say -0.022 m.

Thus 22 mm must be added to each measured length, i.e. $c = -k = 0.022$ m.

Note that by applying equation 3.55,

(a) $$C = \frac{80.473 - (29.972 + 50.475)}{(2-1)} = 0.026 \text{ m}$$

(b) $$C = \frac{210.820 - (29.972 + 50.475 + 130.330)}{(3-1)} = 0.022 \text{ m}$$

3.5.4 The reduction of the slope length to the horizontal

Most EDM instruments measure the slope length (D), and the horizontal length (H) must be derived by computation using either (a) station levels or (b) angular observations.

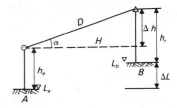

Fig. 3.11

(a) *By station levels* (see Fig. 3.11)
Given the level of station $A = L_a$ (the instrument station), the level of station $B = L_b$ (the reflector station), the height of the EDM $= h_e$, and the height of the reflector $= h_r$, then

$$\Delta h = (L_b + h_r) - (L_a + h_e) = (L_b - L_a) + (h_r - h_e)$$
$$= \Delta L + (h_r - h_e) \quad (3.56)$$

The horizontal length is given by

$$H = (D - c) = \sqrt{(D^2 - \Delta h^2)} \quad (3.57)$$

and the correction is given by

$$c \simeq \Delta h^2/2D + \Delta h^4/8D^3 \quad (3.58)$$

Applying the calculus to the errors gives, by the partial

differentiation of the equation $H^2 = D^2 - \Delta h^2$,

$$2H\delta H = -2\Delta h \delta(\Delta h)$$
$$\delta(\Delta h) = -H\delta H/\Delta h \qquad (3.59)$$

or, by the partial differentiation of the equation $c = \Delta h^2/2D$,

$$\delta c = 2\Delta h \delta(\Delta h)/2D$$
$$\delta(\Delta h) = \Delta h \delta c/2c \qquad (3.60)$$

Example 3.10 If $D = 300.000$ m, $L_a = 31.362$ m AOD, $L_b = 34.475$ m AOD, $h_e = 1.562$ m and $h_r = 1.347$ m, calculate the horizontal length and the limits of error in the levels to give a reduced length to ± 0.001 m.

By equation 3.55,

$$H = \sqrt{(D^2 - \Delta h^2)} = \sqrt{(300.000^2 - 2.898^2)}$$
$$= 299.986 \text{ m}$$

By equation 3.59,

$$\delta(\Delta h) = -H\delta H/\Delta h = -299.986 \times \pm 0.001/2.898$$
$$= \pm 0.1035 \text{ m}$$

Alternatively, by equation 3.58,

$$c = \Delta h^2/2D = 2.898^2/600 = 0.014 \text{ m}$$
$$H = D - c = 300.000 - 0.014 = 299.986 \text{ m}$$

By equation 3.60,

$$\delta(\Delta h) = \Delta h \delta c/2c = 2.898 \times \pm 0.001/(2 \times 0.014)$$
$$= \pm 0.1035 \text{ m}$$

Check. If $\Delta h = 2.898 - 0.104 = 2.794$ then
$$c = 2.794^2/600 = 0.013 \text{ m}$$

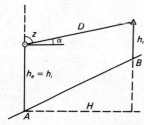

Fig. 3.12

(b) *By angular observations* (see Fig. 3.12)
(i) If $h_e = h_i$ then

$$H = D\cos\alpha = D\sin z \qquad (3.61)$$

where z is the zenith angle, h_i is the height if the theodolite and h_e is the height of the EDM.

(ii) If $h_e \neq h_i$ then (see Fig. 3.13)

$$\alpha = \theta + \delta\theta \text{ (Fig. 3.13)}$$

where θ is the observed vertical angle and

$$\delta\theta = \sin^{-1}\left(\frac{(h_i - h_e)\cos\theta}{D}\right) \qquad (3.62)$$

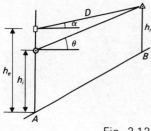

Fig. 3.13

$$\delta\theta° \simeq \frac{57.2958(h_i - h_e)\cos\theta}{D} \qquad (3.63)$$

$$\delta\theta'' \simeq \frac{206\,265\,(h_i - h_e)\cos\theta}{D} \qquad (3.63a)$$

Example 3.11 Given $D = 300.000$ m, $z = 80°\,30'\,20''$, $h_i = 1.462$ m, $h_e = 1.562$ m. Then, by equation 3.63,

$$\delta\theta° = 57.2958° \,(1.462 - 1.562)\sin 80.5056°/300.000$$
$$= -0.0188°$$
$$\alpha = (90° - 80.5056°) - 0.0188° = 9.4756°$$
$$H = 300.000 \cos 9.4756° = 295.907 \text{ m}$$

Applying the calculus to the errors as above, if

$$H = D\cos\alpha = D\sin z$$
then $\quad \delta H = D\cos z\,\delta z = H\cot z\,\delta z \qquad (3.64)$

Applied to the above example,

$$\delta H = 300.000 \cos 80.5056° \times 0.0188/57.2958 = 0.016 \text{ m}$$
$$= 295.975 \cot 80.5056° \times 0.0188/57.2958 = 0.016 \text{ m}$$

If the eccentricity had been ignored,

$$H' = 300.000 \sin 80.5056° = 295.891 \text{ m}$$
$$\text{error} = \quad 0.016 \text{ m}$$

(iii) Where the EDM is attached to the theodolite telescope (see Fig. 3.14)

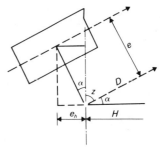

Fig. 3.14

(a) If the vertical angle α is read simultaneously with the pointing of the EDM, the error due to eccentricity (e) is given by

$$e_h = e\sin\alpha \qquad (3.65)$$
and $\quad H = D\cos\alpha - e\sin\alpha \qquad (3.66)$

Note that where the EDM is 0.115 m above the theodolite axis the value of $e\sin\alpha \simeq 2$ mm per degree.

(b) If the theodolite is repointed to the reflector (see Fig. 3.15) then

$AB = D$ (the measured distance)
$AT = e$ (the eccentricity)
$\alpha = $ the angle of inclination of the EDM
$\theta = $ the angle of inclination of the theodolite when pointing to the reflector.

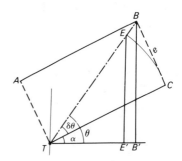

Fig. 3.15

Then

$$TB = \sqrt{(e^2 + D^2)}$$
$$TB' = TB\cos\theta = \sqrt{(e^2 + D^2)}\cos\theta$$
$$TE' = D\cos\theta$$

$$E'B' = TB' - TE' = \sqrt{(e^2 + D^2)} \cos\theta - D\cos\theta$$
$$= \cos\theta[(e^2 + D^2)^{1/2} - D]$$
$$= \cos\theta \frac{(e^2 + D^2 - D^2)}{(e^2 + D^2)^{1/2} + D}$$
$$= e^2 \cos\theta / [(e^2 + D^2)^{1/2} + D]$$

The error in the repointing is approximately

$$(e^2 \cos\theta)/2D \tag{3.67}$$

i.e. $H = D\cos\alpha - e\sin\alpha \simeq D\cos\theta$

Note that when D is greater than 6 m, $E'B'$ is less than 1 mm.

Example 3.12 If $D = 60.000$ m, $e = 0.115$ m, $\alpha = 6°00'00''$, then

$$H = D\cos\alpha - e\sin\alpha$$
$$= 60.000 \cos 6° - 0.115 \sin 6°$$
$$= 59.671 - 0.012 = 59.659 \text{ m}$$

By repointing,

$$H = D\cos\theta$$

where $\theta = \alpha + \delta\theta$.

In the example, $\delta\theta = \tan^{-1}(0.115/60.000) = 0.1098°$. Then $\theta = 6.1098°$ and $H = 60.000 \cos 6.1098° = 59.659$ m as before.

(c) *The reflector is allowed to tilt normal to the line of sight of the telescope* (Fig. 3.16)

Here the theodolite is pointed to the centre of the target below the reflector and thus the EDM is eccentric by the same amount as the reflector. Then

$$H = D\cos\alpha$$

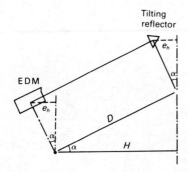

Fig. 3.16

3.6 The reduction of short and medium range EDM slope distances to the spheroid

For short range measurements involving the slope length (D) and the vertical angle α (Fig. 3.17) the horizontal length is given as

$$H = D\cos\alpha$$

As the lengths become longer a better approximation is obtained by assuming that a mean horizontal length H' (Fig. 3.17) is used in the equation, i.e.

$$H' = D\cos\alpha \tag{3.68}$$

The chord length (S') at mean sea level (MSL) relative to this mean horizontal length (H') at a mean level L_m above MSL is

3.6 REDUCTION OF SHORT AND MEDIUM RANGE EDM SLOPE DISTANCES

given by proportion as

$$S'/H' = R/(R + L_m)$$

then $\quad S' = H'R/(R + L_m) \quad$ (3.69)

Written as a 'correction',

$$c = H' - S' = H'[1 - R/(R + L_m)] \quad (3.70)$$
$$\simeq H'L_m/R \quad (3.71)$$

The arc length $(S) = R/\psi_{rad}$ and $\psi/2 = S/2R$. The chord length $(S') = 2R\sin(\psi/2)$ and thus

$$\sin(\psi/2) = S'/2R \quad (3.72)$$

Substituting θ for $\psi/2$ and expanding $\sin^{-1}\theta$ in terms of θ (see Appendix 3) gives

$$\sin^{-1}\theta = S'/2R + (S')^3/48R^3 + 3(S')^5/1280R^5 + \ldots \text{etc.}$$

Then

$$S = 2R[S'/2R + (S')^3/48R^3 + 3(S')^5/1280R^5 + \ldots \text{etc.}]$$
$$= S' + (S')^3/24R^2 + 3(S')^5/640R^4 + \ldots \text{etc.}$$
$$\simeq S'[1 + (S')^2/24R^2] \quad (3.73)$$

Then $c = S - S'$
$$= (S')^2/24R^2 \quad (3.74)$$

Where the levels of stations A and B are known as L_a and L_b then S' may be derived as follows. In triangle ABO of Fig. 3.17,

$$\cos\psi = \frac{(R+L_a)^2 + (R+L_b)^2 - D^2}{2(R+L_a)(R+L_b)}$$

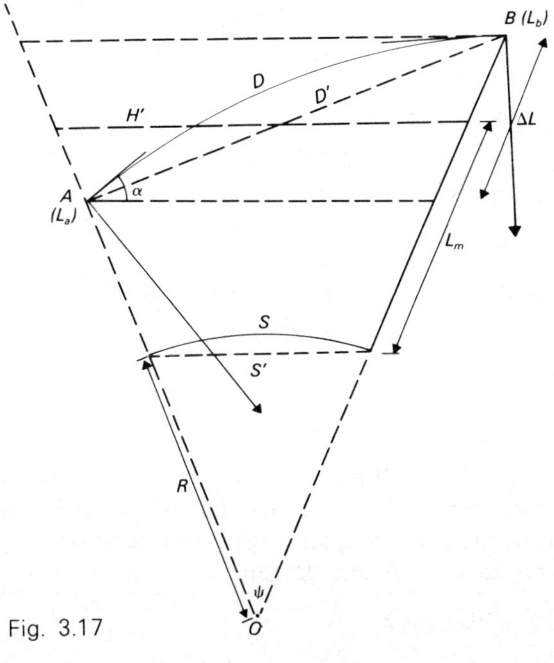

Fig. 3.17

But $\cos\psi = 1 - 2\sin^2(\psi/2)$

Then by equation 3.72,

$$1 - (S')^2/2R^2 = \frac{(R+L_a)^2 + (R+L_b)^2 - D^2}{2(R+L_a)(R+L_b)}$$

So $S' = R\left(\frac{(D-\Delta L)(D+\Delta L)}{(R+L_a)(R+L_b)}\right)^{1/2}$ (3.75)

Example 3.13 Given the measured length 4562.375 m, level of the EDM station 35.460 m AOD, level of the reflector station 152.468 m AOD, calculate the spheroidal length. (Assume the radius of the earth $R = 6370.6$ km.)

$\Delta L = 152.468 - 35.460 = 117.008$ m

By equation 3.75,

$$S' = 6370\,600\left(\frac{(4562.375 - 117.008)(4562.375 + 117.008)}{(6370\,635 \times 6370\,752)}\right)^{1/2}$$

$= 4560.807$ m

By equation 3.74,

$S = 4560.807 + 4560.807^2/(24 \times 6370\,600^2)$

$= 4560.807$ m

Thus at this distance the arc and the chord are the same.

By plane triangle methods,

$H' \simeq \sqrt{(D^2 - \Delta L^2)} = [(D - \Delta L)(D + \Delta L)]^{1/2}$

$= (4562.375^2 - 117.008^2)^{1/2}$

$= 4560.874$

or $H' \simeq D - c$

where $c = 117.008^2/(2 \times 4562.375) = 1.500$

$H' \simeq 4562.375 - 1.500 = 4560.875$ m

By equation 3.69,

$S = H'R/(R+L_m) = 4560.875 \times 6370\,600/6370\,694$

$= 4560.808$ m

If $\alpha = 1°\,28'\,11''$ then

$H' = D\cos\alpha = 4562.375 \cos 1.4697° = 4560.874$ m

In the latter case Helliwell points out that the angle α is not strictly the correct value due to the deviation of the vertical, in azimuth, of the measured line, although if reciprocal observations had been taken as θ_a and θ_b then

$\alpha + \delta\alpha = (\theta_a + \theta_b)/2$

If $H' = D\cos\alpha$ by partial differentiation and writing as errors,

$$\delta H' = -D\sin\alpha\,\delta\alpha \quad \text{and} \quad \delta H'/H' = -\tan\alpha\,\delta\alpha \quad (3.76)$$

Similarly, as

$$H' = (D^2 - \Delta L^2)^{1/2}$$
$$\delta H' = -\Delta L\,\delta(\Delta L)/(D^2 - \Delta L^2)^{1/2}$$

and as

$$\delta H' = D\,\delta\alpha_{rad}$$
$$\delta H'/H' = -D\Delta L\,\delta\alpha/(D^2 - \Delta L^2)^{1/2} \quad (3.77)$$

then

$$\tan\alpha \simeq -D\Delta L/(D^2 - \Delta L^2)^{1/2}$$

Based upon Robbins' observations, Helliwell give the following results.

$\alpha°$	$\delta H'/H'$
3	2 ppm
10	6.8 ppm
20	14.1 ppm

If α is less than 1.5°, the error ratio is less than 1 ppm but when α is greater than 10, the effect is very significant and consideration of the deviation from the vertical (magnitude and direction) is necessary.

Note that if the deviation for the vertical appropriate to the above example is $\delta x = 6''$, then by equation 3.76,

$$\delta H'/H' = -\tan\alpha\,\delta\alpha = -\tan 1.4697 \times 6/206\,265$$
$$= 7.5 \times 10^{-7}$$

and by equation 3.77,

$$= -D\Delta L\,\delta\alpha/(D^2 - \Delta L^2) = 7.5 \times 10^{-7}$$

$\delta H'$ is then $7.5 \times 10^{-7} \times 4560 = 0.003$ m.

3.7 Curvature of the path of the EDM wave

It is generally assumed that for practical purposes the curvature of the path of the wave (σ) is constant over most distances measured in engineering surveying by EDM with wavelengths in or near the visible spectrum (see Fig. 3.17). Thus the coefficient of refraction is given as

$$k = R/2\sigma \quad (3.78)$$

where R is the radius of the spheroid.

The value of σ is dependent upon the vertical temperature and pressure gradients. Given $R = 6.37 \times 10^6$ m and $\sigma = 4.30 \times 10^7$ m then

$$k_1 = 6.37/(2 \times 43.0) = 0.074$$

This represents a typical value for visible light, whilst the value for microwaves based upon $\sigma = 2.0 \times 10^7$ gives

$$k_m = 6.37/(2 \times 20) = 0.159$$

With the observed distance D, the atmospheric adjustment is given as

$$D' = Dn_a/n_{gt} \tag{3.79}$$

where n_a is the standard refractive index (see p. 122) and n_{gt} is the refractive index at the time of measurement. The correction for curvature is then given by Cooper as

$$c = D - D_1 = -D^3 k^2/6R^2 \tag{3.80}$$

The combined effect is given as

$$D_1 = D[n/n - (kD)^2]/6R^2 \tag{3.81}$$

Saastamoinen gives a general equation for velocity and curvature as

$$c = (1 - K)^2 D^3/24 R^2 \tag{3.82}$$

and substituting a value $K = 0.20$ for light waves gives

$$c = D^3/38 R^2 \tag{3.83}$$

and for radio waves $K = 0.25$ gives

$$c = D^3/43 R^2 \tag{3.84}$$

The reduction equations then become, for light waves

$$D_1 = D(1 - D^2/38 R^2) \tag{3.85}$$

and, for radio waves,

$$D_1 = D(1 - D^2/43 R^2) \tag{3.86}$$

Note that in the above equations the coefficient of refraction is not defined in the same manner; $K = R/\sigma$, i.e. $K = 2k$, but even so, there is some discrepancy in the values assumed.

Hodges suggests that reciprocal observations are very desirable and that the coefficient of refraction (K) may be determined as

$$K = 1 - (z_a + z_b - \pi) R/S' \tag{3.87}$$

where z_a and z_b are the zenith angles (in radians) measured at A and B.

From his observations involving lines less than 10 km, corrections are less than 1 mm and are thus negligible for most engineering surveying purposes.

Considering all the factors discussed here, the final spheroidal length (S) is given as

$$S = D + C_1 + C_2 + C_3 \qquad (3.88)$$

where D = the record length. By equations 3.83 and 3.84,

$$C_1 = -D^3/38\,R^2 \quad \text{for light waves}$$
$$\text{or} \qquad = -D^3/43\,R^2 \quad \text{for radio waves}$$
$$C_2 = -(\Delta L^2/2D + \Delta L^4/8D^3)$$
$$\text{or} \qquad = -D(L_a + L_b)2R$$
$$C_3 = (S')^3/24\,R^2 = (D + C_1 + C_2)^3/24\,R^2$$

Example 3.14 Given $D = 9364.360$ m (measured with light waves), $L_a = 30.36$ m AOD, $L_b = 150.42$ m AOD and $R = 6.375 \times 10^6$ m, then

$$C_1 = -9364.360^3/(38 \times 6375\,000^2) = -0.0005 \text{ m}$$

$$C_2 = \frac{-9364.360\,(30.36 + 150.42)}{2 \times 6.375 \times 10^6} = -0.1328 \text{ m}$$

$$C_3 = \frac{(9364.360 - 0.1333)^3}{24 \times 6375\,000^2} = 0.0008 \text{ m}$$

$$S = 9364.360 - 0.133 = 9364.227 \text{ m}$$

It can be seen from the above that C_1 and C_3 are approximately equal and opposite and are thus capable of being neglected for distances less than 10 km.

Exercises 3.2

1 An EDM instrument is thought to have a 'zero' error and to test this the following measurements were made on a multi-pillar base line: $AB = 30.386$ m, $AC = 80.499$ m, $AD = 180.788$ m, $BC = 50.139$ m, $BD = 150.424$ m, $CD = 100.312$ m. Calculate the most probable values of the distances between the pillars and the 'zero' error.

(RICS/M Ans. 30.362 m, 50.114 m, 100.287 m, 0.025 m.)

2 A line AB was measured with an EDM instrument using radio waves and its mean recorded length was 5367.412 m. The manufacturers' specification quotes the standard error as ± 10 mm ± 3 ppm. As a check another measurement was made with an instrument using light waves (specification ± 5 mm ± 1 ppm) and recorded as 5367.435 m. The level of station A was 136.475 m AOD with the instrument height 1.460 m and the level of station B was 169.485 m AOD with the reflector height 1.740 m.

(a) Explain the essential difference between the two types of instruments.

(b) Given the Grid coordinates of B as 364 257.42 m E, 274 692.37 m N, compute the most probable value of the Grid coordinates of A, if the Grid bearing of AB was 326° 27' 40". (Assume $R = 6.37 \times 10^6$ m.)

(TP Ans. 367 221.68 m E, 270 220.46 m N.)

3 As part of a motorway control survey, the following information was derived.

Grid coordinates of the fixed stations:

	E(m)	N(m)
A	473 462.73	399 542.76
B	475 975.18	397 482.38

EDM distance $AC = 4742.375$ m
Height of EDM instrument at $A = 1.57$ m
Height of theodolite at $A = 1.33$ m
Vertical angle to $C = 5° 36' 20"$
Horizontal angle $CAB = 57° 36' 12"$
Calculate the Grid coordinates of station C to the same accuracy as A and B. (TP Ans. 477 943.59 m E, 401 020.26 m N)

4 An EDM instrument attached to a theodolite is to be used to set out a point P from a coordinated baseline AB. It is thought that the use of one face only is necessary for the horizontal angles and tests show that the combined instrumental and observational error is $\pm 01' 00"$. Given the recorded data, calculate the coordinates of P and their standard errors.

Recorded data:

	E(m)	N(m)
A	1 000.000	1 000.000
B	1 236.432	1 542.386

Theodolite station at A
Pointing to B Horizontal circle reading   10° 30' 00"
Pointing to P Horizontal circle reading   25° 45' 00"
                 Vertical zenith angle   86° 25' 45"
Measured slope length $AP = 240.818$ m
(RICS/M Ans. 1150.61 $\pm$ 0.06 m E, 1187.31 $\pm$ 0.04 m N)

5 A and B (OS Grid coordinates 446 273.365 m E, 387 496.327 m N and 446 794.427 m E, 387 004.415 m N respectively) are control points for the setting out of a major structure. From these stations EDM lengths were measured to a third station C, which lies to the north of the line AB, as follows.

AC measured length 963.526 m
    mean vertical angle 3° 26′ 20″
BC measured length 842.697 m
    mean vertical angle 0° 01′ 00″

Compute (a) the angle CAB and (b) the Grid coordinates of C assuming the angle ABC is 75° 40′ 27″ and the angles are thought more reliable than the lengths.
        (TP Ans. 58° 05′ 43″, 447 203.16 m E, 387 741.01 m N)

6   A line AB was measured with an EDM instrument G/12 and then with another instrument G/6 and the following data was recorded
   Length with the G/12   69.190 m ± 3 mm
   (reflector constant −0.025 m)
   Length with the G/6   69.299 m ± 7 mm
   (reflector constant −0.138 m)
   Height of the theodolite at A   1.423 m
   Level of A   27.362 m AOD
   Height of G/12   1.538 m
   Height of G/6   1.558 m
   Height of reflector at B   1.367 m

The theodolite was pointed at the base of the reflector 115 m below its centre when the slope length was recorded with the G/12 and a vertical angle of 3° 26′ 32″ was recorded. It was then repointed to the centre of the reflector and 3° 32′ 10″ was recorded. Compute the most probable value of the horizontal length of the line AB and the level of B showing that the data given is compatible.   (TP Ans. 69.034 m, 31.684 m AOD)

7   An EDM (using light waves) measured a length AB as 5672.368 m. The heights of A and B were 126.37 m and 382.46 m AOD respectively. Calculate the spheroidal length, assuming $n = 1.000\,330$ and $K = 0.07$, $R = 6.37 \times 10^6$ m.
        (TP Ans. 5666.34 m)

3.8 Trigonometrical heighting

Generally, trigonometrical heighting is far less accurate than precise spirit levelling but within a triangulation network, where a levelling control network exists, this method may be used to give adequate height control for topographical mapping.

3.8.1 Applied to plane surveying

Here the method consists of the measurement of the slope length (D) or the horizontal length (H) together with the vertical angle (α).

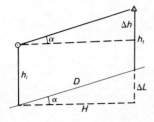

Fig. 1.18

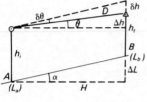

Fig. 3.19

(i) When the instrument height (h_i) and the target height (h_t) are the same (Fig. 3.18), then

$$\Delta h = \Delta L = D \sin \alpha = H \tan \alpha \qquad (3.89)$$

(ii) When $h_i \neq h_t$ (Fig. 3.19) then

$$\Delta h = H \tan \theta \qquad (3.90)$$

or $\quad \Delta L = D' \sin \alpha = \Delta h + h_i - h_t \qquad (3.91)$

This difference between h_i and h_t ($= \delta h$), may be transformed into an angular adjustment known as the 'eye and object' adjustment ($\delta \theta$). Then

$$\alpha = \theta + \delta \theta$$

where $\quad \delta \theta = \sin^{-1} \dfrac{(h_i - h_t) \cos \theta}{D} = \sin^{-1} (\delta h \cos \theta)/D \qquad (3.92)$

$$\delta \theta° = 57.2958 (\delta h \cos \theta)/D \qquad (3.93)$$

or $\quad \delta \theta'' = 206\,265 (\delta h \cos \theta)/D \qquad (3.94)$

where D is the EDM measurement along the line of slope (θ) and D' is the length of the line of slope (α). The level of station $B(L_b)$, relative to the level of station $A(L_a)$, is given as

$$L_b = L_a + \Delta L = L_a + \Delta h + h_i - h_t$$
$$= L_a + \Delta h + \delta h \qquad (3.95)$$

Example 3.15 Given $L_a = 125.36\,\text{m}$, $\theta = 1°\,10'\,30.0''$, $h_i = 1.45\,\text{m}$, $h_t = 4.78\,\text{m}$, $D = 6372.65\,\text{m}$.

$$H = D \cos \theta = 6372.65 \cos 1.1750° = 6371.31\,\text{m}$$
$$\Delta h = D \sin \theta = 6372.65 \sin 1.1750° = 130.68\,\text{m}$$
$$L_b = L_a + \Delta h + h_i - h_t$$
$$= 125.36 + 130.68 + 1.45 - 4.78 = 252.71\,\text{m}$$

Alternatively,

$$\delta \theta = 57.2958 (\delta h \cos \theta)/D$$
$$= 57.2958 \times (-3.33) \cos (1.1750°)/6372.65$$
$$= -0.0299° = 107.64''$$
$$\Delta L = H \tan \alpha = 6371.31 \tan (1.1750 - 0.0299)°$$
$$= 127.35\,\text{m}$$
$$L_b = L_a + \Delta L = 125.36 + 127.35 = 252.71\,\text{m}$$

Note that in Clark, Vol II, the 'eye and object' adjustment is given as

$$\delta \theta'' = \delta h \cos^2 \theta / (D \sin 1'') \qquad (3.96)$$

Using the above values gives $\delta \theta'' = -107.74''$; cf. $107.64''$. The extra $\cos \theta$ therefore does not appear necessary for most engineering surveying purposes.

3.8.2 Applied to geodetic surveying

The linear measurements may be by EDM along the line of sight or may be derived from the grid coordinates, when the grid length would need to be converted into a spheroidal length by the application of a scale factor. Applied to the UK the local scale factor (F) is given as

$$F = 0.999\,601\,27(1 + Y^2 \times 1.228 \times 10^{-8})$$

where Y is the distance from the central meridian in km (see p. 178). When the lengths become greater than, say, 6 km, the effect of the earth's curvature becomes significant and also the question of atmospheric refraction must be considered.

3.8.3 Curvature and refraction

1) *'Correction' for curvature* ($c/2$)

In Fig. 3.20 let A and B be at the same height above MSL i.e. $L_a = L_b$. The effect of the earth's curvature can be seen to produce an error in the vertical angle and an angular adjustment ($c/2$) is necessary before computing Δh. If c is the angle subtended at the centre of the earth by the spheroidal arc S (at MSL) then

$$S = Rc \qquad (3.97)$$

and $\quad c = S/R \qquad (3.98)$

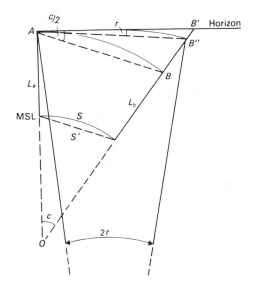

Fig. 3.20

Applied to the vertical angle,

$$(c/2)'' = 206\,265\,S/2R \qquad (3.99)$$

The error in the height of B due to $(c/2)$ is given as

$$BB' \simeq Sc/2 = \delta L_c = S^2/2R \qquad (3.100)$$

2) *'Correction' for refraction* (r)
The light path is refracted by atmospheric density variations. The coefficient of vertical refraction (k) is defined as the ratio (r/c), i.e.

$$k = r/c \qquad (3.101)$$
Then $r = kc$ $\qquad (3.102)$

The average values of k used in practice are UK (over land), 0.075; UK (over water), 0.081; South Africa, 0.055; USA, 0.065.

3) *The combination of curvature and refraction* (e)
The 'correction' for curvature is positive and the 'correction' for refraction is negative. As the former is always larger, then the combined effect is given as

$$e = c/2 - r \qquad (3.103)$$
$$= S/2R - kS/R = S(1 - 2k)/2R \qquad (3.104)$$

The error in height due to this combination (e) is given as

$$\delta L_e = S^2(1 - 2k)/2R \qquad (3.105)$$

Note that

$$k = 0.5 - Re/S \qquad (3.106)$$

Example 3.16 Given $S = 1000.00$ m, $R = 6.37 \times 10^6$ m, $k = 0.07$, then by equation 3.104,

$$e'' = 206\,265 \times 1000.00(1 - 0.14)/(2 \times 6\,370\,000) = 13.9''$$

and by equation 3.78,

$$c'' = 206\,265 \times 1000.00/6\,370\,000 = 32.4''$$
$$r'' = c/2'' - e'' = 16.2'' - 13.9'' \quad = 2.3''$$

Check. $k = r/c = 2.3/32.4 \qquad = 0.07$

3.8.4 To obtain the level of station B (L_b)

With reference to Fig. 3.21,

$$L_b = L_a + S \tan \theta (1 + L_m/R) + S^2(1 - 2k)/2R + \delta h \qquad (3.107)$$

The approximate value of L_b must be obtained using $\Delta h' = S \tan \theta$, i.e. $L_b' = L_a + \Delta h'$.

Then $L_m = (L_a + L_b')/2$

If the 'eye-object' adjustment is made, then

$$L_b = L_a + S \tan \theta'(1 + L_m/R) + S^2(1 - 2k)/2R \qquad (3.108)$$

Note that SL_m/R is the adjustment to the length for the mean

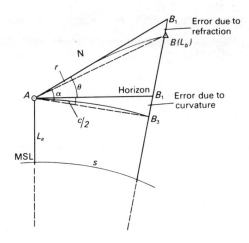

Fig. 3.21

height L_m, and that, $S^2(1-2k)/2R$ is the adjustment for curvature and refraction.

Example 3.17 Given the spheroidal length $AB = 3480.8$ m, $h_i = 1.43$ m, $h_t = 3.90$ m, $\theta = 1°49'05.0''$, $L_a = 128.38$ m, $R = 6.369 \times 10^6$ m and $k = 0.074$, then

$$L'_b = L_a + S\tan\theta = 128.38 + 3480.8 \tan 1.8181°$$
$$= 128.38 + 110.489 = 238.87 \text{ m}$$
$$L_m = (128.38 + 238.87)/2 = 183.63 \text{ m}$$

By equation 3.107,

$$L_b = 128.38 + 110.489(1 + 0.000\,029)$$
$$+ 3480.8^2(1 - 0.148)/(2 \times 6.369 \times 10^6)$$
$$+ 1.43 - 3.90$$
$$= 128.38 + 110.49 + 0.81 - 2.47 = 237.21 \text{ m}$$

Alternatively, by the 'eye-object' adjustment,

$$\delta\theta° = 57.2958\,(1.43 - 3.90)\cos 1.8181°/3480.8$$
$$= -0.0406° = -146.3''$$
$$\theta' = 1.8181° - 0.0406° = 1.7775°$$

then $\quad L_b = 128.38 + 108.02 + 0.81 \quad = 237.21 \text{ m}$

The curvature and refraction values are then, by equation 3.104,
$e'' = 206\,265 \times 3480.8(1 - 0.148)/(2 \times 6.369 \times 10^6) = 48.02''$
by equation 3.98,
$c'' = 206\,265 \times 3480.8/6.369 \times 10^6 \qquad = 112.73''$
by equation 3.102,
$r'' = 0.074 \times 112.73 \qquad\qquad\qquad\qquad = 8.34''$

Example 3.18 If observations from B were taken over the same line and the results recorded as $\theta_b = -1°45'18.0''$, $h_i = 1.50$ m, $h_t = 4.48$ m, then

$$S \tan \theta_b (1 + L_m/R) = -3480.8 \tan 1.7550°(1.000\,029)$$
$$= -106.66 \text{ m}$$
$$L_b = 128.38 - (-106.66 + 0.81 + 1.50 - 4.48)$$
$$= 237.21 \text{ m}$$

Alternatively,

$$\delta\theta = 57.2958(1.50 - 4.48)/3480.8 = -0.0491°$$
$$\theta = -1.7550° - 0.0491° = -1.8041°$$

Then $L_b = 128.38 - (-109.64 + 0.81) = 237.21$ m

3.8.4 Reciprocal observations

The vertical angles θ_a and θ_b are the angles between the horizon and the tangents to the light paths. When the 'eye-object' adjustments are made, i.e.

$$\theta_a = (\theta_0 + \delta\theta)_a \quad \text{and} \quad \theta_b = (\theta_0 + \delta\theta)_b$$

these adjusted values would be equal were they not affected by curvature and refraction. Note that it is mathematically convenient to consider here the zenith angles (frequently recorded on modern surveying instruments) where $Z = (90 - \theta)$ and thus $\theta = (90 - Z)$. Two cases are possible:

1) Δh is very large and thus one of the angles (θ) is an elevation and the other is a depression;
2) Δh is relatively small, when both observations are depressions.

In both cases the following applies (see Figs 3.22 and 3.23).

At A $\quad \alpha_a = (90 - Z_a) + c_a/2 - r_a$
At B $\quad \alpha_b = (Z_b - 90) - c_b/2 + r_b$
$\quad -\alpha_b = (90 - Z_b) + c_b/2 - r_b$

If the observations are made simultaneously, then assuming the light path to be circular, and the refraction to be constant, and if $c_a = c_b = c$ and $r_a = r_b = r$,

$$\alpha = (Z_b - Z_a)/2 \qquad (3.109)$$
$$= [(90 - \theta_b) - (90 - \theta_a)]/2 = (\theta_a - \theta_b)/2 \qquad (3.110)$$

and $\quad 0 = 180 - (Z_a + Z_b) + c - 2r$

Then $\quad r = 90 - (Z_a + Z_b)/2 + c/2 \qquad (3.111)$
$$= 90 - [(90 - \theta_a) + (90 - \theta_b)]/2 + c/2$$
$$= (\theta_a + \theta_b)/2 + c/2 \qquad (3.112)$$

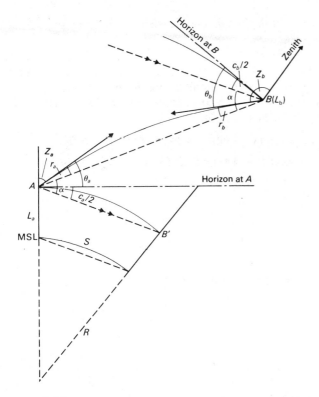

Fig. 3.22

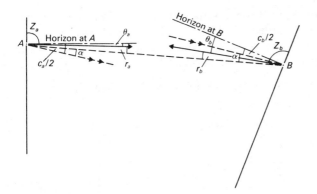

Fig. 3.23

The reduced height of B relative to A is given as

$$L_b = L_a + S \tan \alpha (1 + L_m/R) \tag{3.113}$$

Note that Clark Vol. II gives the more precise equation

$$\Delta L = S \tan \alpha [1 + L_m/R + S^2/(12R^2)] \tag{3.114}$$

The term $S^2/12R^2$ is usually very small and represents an adjustment for the fact that in the triangle ABB', the angle B' is not $90°$.

Example 3.19 Assume that the results in Examples 3.18 and 3.19 were obtained by simultaneous reciprocal observations (note that $Z_a = 88° 10' 55.0''$ and $Z_b = 91° 45' 18.0''$). The eye-object adjustments become

$$\delta\theta_a = (1.43 - 3.90) \times 57.2958/3480.8 = -0.0407°$$
$$\delta\theta_b = (1.50 - 4.48) \times 57.2958/3480.8 = -0.0491°$$

Then $\quad \theta_a = 1.8181 - 0.0407 = 1.7774°$

or $\quad Z_a = 88.2226°$

$\quad \theta_b = -1.7550 - 0.0491 = -1.8041°$

or $\quad Z_b = 91.8041°$

Then $\quad \alpha = (1.7774 + 1.8041)/2$
$$= (91.8041 - 88.2226)/2 = 1.7904°$$

and $\quad L_b = L_a + S \tan \alpha \, (1 + L_m/R)$
$$= 128.38 + 3480.8 \tan 1.7904 (1.000\,029)$$
$$= 237.21 \text{ m}$$

as before

$$r = [(1.7774 - 1.8041) + (57.2958 \times 3480.8/6.369 \times 10)]/2$$
$$= 0.0023 = 8.30''$$

Example 3.20 From a station A to a station B the following data was recorded: VA $90° 01' 44.0''$; $h_i = 1.65$ m; $h_t = 2.29$ m; $L_a = 1168.21$ m; $AB = 7933.6$ m (assume $R = 6.37 \times 10^6$ m and $k = 0.056$). Adjusting equation 3.93,

$$\delta Z = 57.2958(2.29 - 1.65) \sin 90° 01' 44.0''/7933.6$$
$$= 0.0046°$$

Then $\quad Z = 90.0289'' + 0.0046 = 90.0335°$

$\quad L_m = L_a + \Delta h/2$
$$= L_a + (S/\tan Z)/2$$
$$= 1168.21 + (7933.6/\tan 90.0335)/2 = 1165.89 \text{ m}$$

Adjusting equation 3.108,

$L_b = L_a + S(1 + L_m/R)/\tan Z + S^2(1 - 2k)/2R$
$$= 1168.21 + 7933.6(1 + 1165.89/6.37 \times 10^6)/\tan 90.0335$$
$$\quad + 7933.6^2(1 - 0.112)/12.74 \times 10^6$$
$$= 1168.21 - 4.640 + 4.387 = 1167.96 \text{ m}$$

Example 3.21 Reciprocal observations made from B to A were recorded as VA $90° 00' 56.0''$; $h_i = 1.52$ m; $h_t = 3.81$ m. As before,

$$\delta Z = 57.2958(3.81 - 1.52)/7933.6 = 0.0165$$
$$Z = 90.0156 + 0.0165 = 90.0321$$

Then $L_b = 1168.21 + 7933.6(1.0002)/\tan 90.0321 + 4.387$
$= 1168.21 + 4.45 - 4.39 = 1168.27 \text{ m}$

Using both sets of observations the mean value is $(1167.96 + 1168.27)/2 = 1168.11$ m.

If the observations had been assumed to have been simultaneous then

$$\alpha = (Z_b - Z_a)/2$$
$$= (90.0321 - 90.0335)/2 = -0.0007$$

Then $L_b = 1168.21 + 7933.6 (1.0002)(\tan -0.0007)$
$= 1168.11 \text{ m}$

From the reciprocal observations the value of refraction must have been assumed incorrectly. Thus

$$c = 57.2958 \times 7933.6/6.37 \times 10^6$$
$$= 0.0714° = 256.89''$$
$$r = 90 - (Z_a + Z_b)/2 + c/2$$
$$= 90 - (90.0335 + 90.0321)/2 + 0.0714/2$$
$$= 0.0029° = 10.44''$$

Then $k = r/c = 0.0029/0.0714$
$= 0.0406$

The effect of curvature and refraction on the level would then be

$$7933.6^2 (1 - 0.081)/(12.74 \times 10^6) = 4.54 \text{ m}$$

Compared with the value of 4.39 using $k = 0.056$ means that an adjustment of $(4.54 - 4.39) = 0.15$ m should be made to each observation, i.e.

From A $1167.96 + 0.15 = 1168.11$ m
From B $1168.27 - 0.15 = 1168.12$ m

Exercises 3.3

1 The horizontal distance between two stations P and Q is 5951.30 m. A theodolite at P sights on to a beacon adjacent to station Q at the same time as a theodolite at Q sights on to a beacon adjacent to station P. The following measurements are obtained: angle of elevation recorded at $P = 01° 19' 38''$; angle of depression recorded at $Q = 01° 21' 01''$; height of beacon at $Q = 2.36$ m; height of instrument at $P = 1.36$ m; height of instrument at $Q = 1.47$ m; height of beacon at $P = 2.85$ m.

Determine the difference in level between the two stations and the coefficient of atmospheric refraction. (Assume the radius of the earth as 6.37×10^6 m.)   (ICE Ans. 139.27 m; 0.070)

2   From station A, a station B, 29 800.3 m distant, is observed with a depression angle of 04′ 08″ and a reciprocal observation from B to A gave a depression angle of 09′ 41″; in both cases the sights were taken to signals at the same height above the ground as the observing theodolite. Calculate (a) the refraction 'correction' (b) the refraction coefficient k, and (c) the difference in level. (Assume 1″ of arc as 30.91 m.)

(Ans. 67.6″, 0.07, 24.06 m)

3   Reciprocal vertical angles were observed between two stations P and Q as follows: mean vertical angle P to Q $+1°08′15″$; mean vertical angle Q to P $-0°53′30″$; heights at P, signal 5.25 m, theodolite 1.10 m; heights at Q, signal 4.00 m, theodolite 1.20 m; geodetic distance PQ 1547 m. Taking the radius of the earth as 6.382×10^6 m, calculate the value of the coefficient of refraction and the difference in elevation of P and Q.   (LU Ans. 0.08, 26.72 m)

Bibliography

AGA Publications
ASHKENAZI, V., and DODSON, A. H., The calibration and evaluation of EDM instruments. *Proc. XV FIG Congress*, Stockholm (1977)
ASHKENAZI, V., and DODSON, A. H., *Measurement of Deformations by Surveying Techniques*. Prepared for Nottingham Seminar (1978)
BARRELL, H., and SEARS, J. E., The refraction and dispersion of air for the visible spectrum. *Phil. Trans. Royal Soc.* (1939)
BURNSIDE, C. D., *Electronic Distance Measurement*. Crosby Lockwood (1971)
CLARK, D., and GLENDINNING, J., *Plane and Geodetic Surveying*, II. Constable (1963)
COOPER, M. A. R., *Fundamentals of Survey Measurement and Analysis*. Crosby Lockwood Staples (1974)
ESSEN, L., and FROOME, K. D., The refractive index of air for radio waves and microwaves. *Proc. Phys. Soc.*, B54 (1951)
HELLIWELL, E. G., Deviation of the vertical and reduction of short range EDM slope distances to the spheroid. *Chartered Surveyor Quarterly*, **5**, 3 (1978)
HODGES, D. J., Arc and beam corrections in EDM. *Institution of Mining Engineers* (December 1969)
HODGES, D. J., *Optical Distance Measurement* (1971)
HODGES, D. J., SKELLERN, P. and MORLEY, J. A., Trials with Model 6 geodimeter for surface surveys. *Institution of Mining Engineers* (1967)
ROBBINS, A. A., A geoidal section through Great Britain. *Survey Review*, XVII, 129 (July 1963)
SAASTAMOINEN, J., Curvature corrections in EDM. *Bulletin Geod*, 73 (September 1964)
SCHWENDENER, H. R., Electronic distancers for short-range accuracy and checking procedures. *Survey Review*, **XII**, 164 (1972)
UREN, J., and PRICE, W. F., *Surveying for Engineers*. Macmillan (1978)

4
Application of spherical trigonometry to engineering surveying

4.1 Definitions

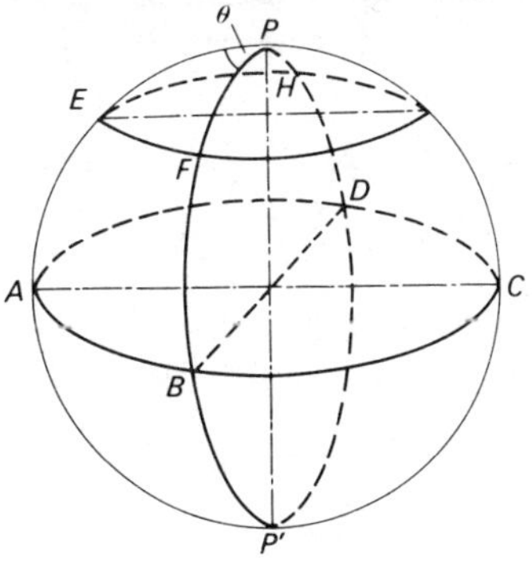

The section of a sphere by an intersecting plane is a circle at its periphery. If the plane passes through the centre of the sphere, the radius of this circle is a maximum and the circle is known as a **great circle** (Fig. 4.1), e.g. $ABCD$ and $BFPDHP'$. Any other intersecting plane produces a **small circle**, e.g. $EFGH$. Excluding the poles, only one great circle can pass through any two given points on the sphere and the arc of this circle is the shortest distance between the points, e.g. $AB = CD$, $BF = HD$. A **spherical triangle** is bounded by the arcs of three great circles which form the sides, and their intersection form the vertices of the triangle, e.g. $\triangle APB$. The **axis** of any circle of a sphere is that diameter which is perpendicular to the plane of the circle; the extremities of the diameter are known as the **poles** of the circle, e.g. PP'.

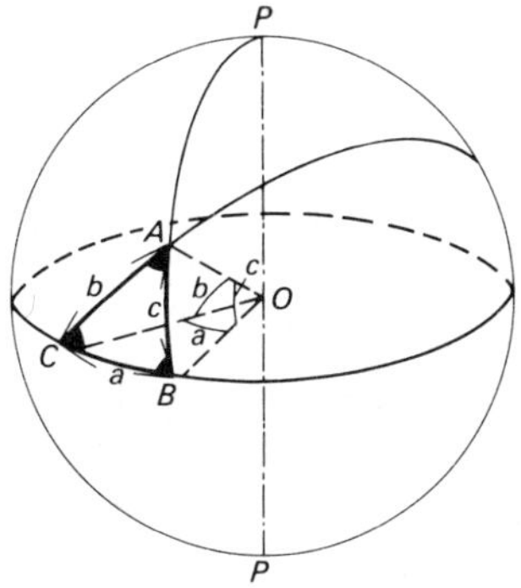

Fig. 4.1

4.1.1 Notation

Spherical triangle ABC has sides a, b and c considered as angles subtended at the centre O, i.e. in Fig. 4.2 $a = BOC$, $b = COA$, $c = BOA$. The angles at the vertices A, B and C are considered as contained in the planes tangential to the sphere at each point and are thus the angle contained between the tangents to the arcs forming the vertex, e.g. A as shown. As each side is assumed to be less than a semicircle and each angle less than $180°$ then $180° < (A+B+C) < 540°$.

Fig. 4.2

4.1.2 The area of the spherical triangle

A **lune** of a sphere is the surface bounded by two great circles, e.g. in Fig. 4.1 $PFBP'AEP$. At the poles the angle θ is subtended and the area is proportional to this angle, i.e.

$$\frac{\text{Area of lune}}{\text{Surface area of the sphere}} = \frac{\theta}{2\pi}$$

$$\text{Area of lune} = 4\pi R^2 \theta / 2\pi = 2R^2\theta \qquad (4.1)$$

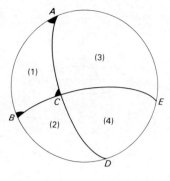

Fig. 4.3

In Fig. 4.3,

$$\text{area of lune } ABDCA = \Delta_1 + \Delta_2 = 2R^2 A$$
$$\text{area of lune } BCEAB = \Delta_1 + \Delta_3 = 2R^2 B$$
$$\text{area of lune } CBAC + CDEC = \Delta_1 + \Delta_4 = 2R^2 C$$

Summing these equations gives

$$\Delta_1 + \Delta_2 + \Delta_3 + \Delta_4 + (2 \times \Delta_1) = 2R^2(A + B + C)$$

But $\Delta_1 + \Delta_2 + \Delta_3 + \Delta_4 = \text{area of hemisphere} = 2\pi R^2$

and $(A + B + C) = \pi + E$

$$2\pi R^2 + (2 \times \Delta_1) = 2\pi R^2 + 2R^2 E$$

Then the spherical excess is given by Δ_1/R^2 radians

$$E = \left(\frac{\text{Area of triangle } ABC}{R^2 \sin 1''} \right) \text{ seconds} \qquad (4.2)$$

Note that on the earth's surface, a spherical triangle of 500 km² has an E value of 2.5″.

Legendre's theorem states that where the spherical excess is small then $E/3$ is deducted from each adjusted spherical angle and plane trigonometry may replace spherical trigonometry, i.e.

$$\frac{a}{\sin(A - E/3)} = \frac{b}{\sin(B - E/3)} = \frac{c}{\sin(C - E/3)} \qquad (4.3)$$

The calculation of the area of the spherical triangle is sufficiently accurate using plane trigonometrical values but L'Huilier's equation is given as

$$E/4 = \tan^{-1}[\tan(s/2)\tan(s-a)/2\tan(s-b)/2\tan(s-c)/2]^{1/2} \qquad (4.4)$$

where $s = (a + b + c)/2$.

Alternatively,

$$E = \frac{c^2 \sin A \sin B}{2 \rho v \sin C} \qquad (4.4a)$$

or

$$E \simeq bc \sin A / 2R^2 \qquad (4.4b)$$

4.2 Spherical trigonometrical formulae

The two most important formulae in spherical trigonometry (as in plane trigonometry) from which all the others may be derived are as follows

1) The sine rule $\qquad \dfrac{\sin a}{\sin A} = \dfrac{\sin b}{\sin B} = \dfrac{\sin c}{\sin C} \qquad (4.5)$

2) The cosine rule $\qquad \cos A = \dfrac{\cos a - \cos b \cos c}{\sin b \sin c} \qquad (4.6)$

4.3 Solution of the general spherical triangle

The similarity between the spherical and plane trigonometrical formulae may be seen in Appendix 2.

There are six elements of the spherical triangle ABC, i.e. three angles A, B and C and three sides a, b and c (see Fig. 4.2). Usually, given three elements, the remaining three may be found.

Case 1
Given three angles A, B and C, then the polar triangle application to the cosine rule (see Section 4.3.1) gives

$$\cos a = \frac{\cos A + \cos B \cos C}{\sin B \sin C} \quad \text{etc.}$$

Example 4.1 Given $A = 63°\,27'\,30''$, $B = 54°\,38'\,27''$, $C = 102°\,17'\,36''$, then

$$\cos a = \frac{\cos 63.4583° + \cos 54.6408° \cos 102.2933°}{\sin 54.6408° \sin 102.2933°}$$

$$a = \cos^{-1} 0.406\,147 = 66.0370° = 66°\,02'\,13''$$

Similarly

$$b = 56.4125° = 56°\,24'\,45''$$
$$\underline{c = 86.4107° = 86°\,24'\,38''}$$
$$208.8602° = 208°\,51'\,36''$$

Case 2
Given three sides a, b and c, then

$$\cos A = \frac{\cos a - \cos b \cos c}{\sin b \sin c} \quad \text{etc.}$$

Example 4.2 Given $a = 66.0370°$, $b = 56.4125°$, $c = 86.4107°$, then

$$\cos A = \frac{\cos 66.0370° - \cos 56.4125° \cos 86.4107°}{\sin 56.4125° \sin 86.4107°}$$

$$A = \cos^{-1} 0.446\,849 = 63.4583$$
$$B = 54.6408$$
$$C = \cos^{-1} -0.210\,874 = 102.2933$$

Case 3
Given two angles (say B and C) and one side opposite one of

them (say b) then
$$\sin C = \sin b \, \frac{\sin c}{\sin B}$$

Note that this has two solutions.
To obtain the remaining elements,
$$\cot A/2 = \frac{\cos\tfrac{1}{2}(b+c)}{\cos\tfrac{1}{2}(b-c)} \tan\tfrac{1}{2}(B+C)$$

then $\sin a = \sin b \sin A / \sin B$

Example 4.3 Given $B = 65.3533°$, $C = 94.4825°$, $b = 55.1050°$, then

$$\sin c = \sin 55.1050° \sin 94.4825° / \sin 65.3533°$$
$$c = \sin^{-1} 0.899\,655 = 64.1127° \quad (\text{or } 115.8873°)$$
$$\cot A/2 = \frac{\cos\tfrac{1}{2}(55.1050° + 64.1127°)}{\cos\tfrac{1}{2}(55.1050° - 64.1127°)} \tan\tfrac{1}{2}(65.3533° + 94.4825°)$$
$$A = 2 \cot^{-1} 2.854\,071 = 38.6185°$$

or $\cot A/2 = \dfrac{\cos\tfrac{1}{2}(55.1050° + 115.8873°)}{\cos\tfrac{1}{2}(55.1050° - 115.8873°)} \tan\tfrac{1}{2}(65.3533° + 94.4825°)$

$$A = 2 \cot^{-1} 0.511\,944 = 125.7756°$$
$$\sin a = \sin 55.1050° \sin 38.6185° / \sin 65.3533°$$
$$a = \sin^{-1} 0.563\,226 = 34.2792°$$

or $\quad \sin a = \sin 55.1050° \sin 125.7756° / \sin 65.3533°$
$$a = \sin^{-1} 0.732\,141 = 47.0662°$$

As $a + b$ must be $> c$,
$$34.2792° + 55.1050° = 89.3842°$$

when $C = 64.1127°$ and when $A = 38.6185°$
$$47.0662° + 55.1050° = 102.1712° \text{ as } C = 115.8873°$$

Therefore the last solution is not applicable.

Case 4
Given two angles (say A and B) and one adjacent side (say c),
$$\tan\tfrac{1}{2}(a+b) = \frac{\cos\tfrac{1}{2}(A-B)}{\cos\tfrac{1}{2}(A+B)} \tan\tfrac{1}{2}c$$

and $\quad \tan\tfrac{1}{2}(a-b) = \dfrac{\sin\tfrac{1}{2}(A-B)}{\sin\tfrac{1}{2}(A+B)} \tan\tfrac{1}{2}c$

and $\quad \cos C = \dfrac{\cos c - \cos a \cos b}{\sin a \sin b}$

4.3 SOLUTION OF THE GENERAL SPHERICAL TRIANGLE

Example 4.4 Given $A = 51.3672°$, $B = 78.3314°$, $c = 48.5547°$, then

$$\tan \tfrac{1}{2}(b+a) = \frac{\cos \tfrac{1}{2}(78.3314° - 51.3672°)}{\cos \tfrac{1}{2}(78.3314° + 51.3672°)} \tan 24.27735°$$

$$\tfrac{1}{2}(b+a) = \tan^{-1} 1.032027 = 45.902969°$$

$$\tan \tfrac{1}{2}(b-a) = \frac{\sin 13.48210°}{\sin 64.84930°} \tan 24.27735°$$

$$\tfrac{1}{2}(b-a) = \tan^{-1} -0.116170 = 6.626364°$$

By adding, $b = 52.5293°$

By subtracting, $a = 39.2766°$

$$\cos C = \frac{\cos 48.5547° - \cos 39.2766° \cos 52.5293°}{\sin 39.2766° \sin 52.5293°}$$

$$C = \cos^{-1} 0.380099 = 67.6602°$$

Case 5
Given two sides (say b, c) and an included angle A, then

$$\cos a = \cos b \cos c + \sin b \sin c \cos A$$

and $$\tan \tfrac{1}{2}(B+C) = \frac{\cos \tfrac{1}{2}(b-c)}{\cos \tfrac{1}{2}(b+c)} \cot \tfrac{1}{2} A$$

$$\tan \tfrac{1}{2}(B-C) = \frac{\sin \tfrac{1}{2}(b-c)}{\sin \tfrac{1}{2}(b+c)} \cot \tfrac{1}{2} A$$

Example 4.5 Given $b = 50.3692°$, $c = 72.1148°$, $A = 41.1987°$, then

$$\cos a = \cos 50.3692° \cos 72.1148°$$
$$+ (\sin 50.3692° \sin 72.1148° \cos 41.1987°)$$
$$a = \cos^{-1} 0.747381 = 41.6360°$$

$$\tan \tfrac{1}{2}(C+B) = \frac{\cos \tfrac{1}{2}(72.1148° - 50.3692°)}{\cos \tfrac{1}{2}(72.1148° + 50.3692°)} \cot 20.59935°$$

$$\tfrac{1}{2}(C+B) = \tan^{-1} 5.430735 = 79.566592°$$

$$\tan \tfrac{1}{2}(C-B) = \frac{\sin 10.87280°}{\sin 61.24200°} \cot 20.59935°$$

$$\tfrac{1}{2}(C-B) = \tan^{-1} 0.572465 = 29.789645°$$

By adding, $C = 109.3562°$
$B = 49.7769°$

Example 4.6 Given $a = 70°\,15'\,00''$, $b = 50°\,20'\,02''$, $c = 40°\,09'\,59''$, then by the cosine rule (equation 4.6),

$A = \cos^{-1}[(\cos a - \cos b \cos c)/(\sin b \sin c)] = 107.5678° = 107°\,34'\,04''$
$B = \cos^{-1}[(\cos b - \cos a \cos c)/(\sin a \sin c)] = 51.2370° = 51°\,14'\,13''$
$C = \cos^{-1}[(\cos c - \cos a \cos b)/(\sin a \sin b)] = \underline{40.7954°} = \underline{40°\,47'\,43''}$

$\phantom{C = \cos^{-1}[(\cos c - \cos a \cos b)/(\sin a \sin b)] = } 199.6002° \quad 199°\,36'\,00''$

$E \phantom{= \cos^{-1}[(\cos c - \cos a \cos b)/(\sin a \sin b)] } = 19.6002° = 19°\,36'\,00''$

Check by the sine rule (equation 4.5),

$B = \sin^{-1}(\sin b \sin C/\sin c) = 51.2370°$ as above

4.3.1 The polar triangle

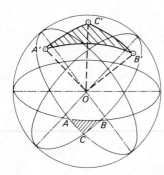

Fig. 4.4

If A', B' and C' are the poles of the arcs of the spherical triangle ABC (Fig. 4.4), i.e. A' is the pole of the arc BC, B' the pole of AC and C' the pole of AB, then the triangle $A'B'C'$ is the polar triangle of the primitive spherical triangle ABC and vice versa. The sine and cosine rules apply but the following relationships should be noted.

$$a' = 180° - A \qquad b' = 180° - B \qquad c' = 180° - C$$

The application of the polar triangle is as follows.

As $\quad \cos a' = \cos b' \cos c' + \sin b' \sin c' \cos A'$

then $\quad \cos(180° - A) = \cos(180° - B)\cos(180° - C)$
$ + \sin(180° - B)\sin(180° - C)\cos(180° - a)$

i.e. $\quad -\cos A = \cos B \cos C - \sin B \sin C \cos a$

and $\quad \cos a = (\cos A - \cos B \cos C)/(\sin B \sin C)$

4.3.2 The right angled spherical triangle

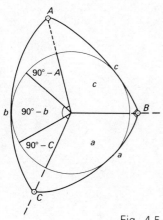

Fig. 4.5

If the angle B is assumed 90° the following ten relationships exist.

$\sin a = \sin b \sin A = \tan c \cot C$ \qquad (4.7)
$\cos b = \cos a \cos c = \cot A \cot C$ \qquad (4.8)
$\sin c = \sin C \sin b = \cot A \tan a$ \qquad (4.9)
$\cos A = \cos a \sin C = \cot b \tan c$ \qquad (4.10)
$\cos C = \cos c \sin A = \tan a \cot b$ \qquad (4.11)

4.3.3 Napier's mnemonics

Neglecting the angle containing the right angle, the circle shown in Fig. 4.5 is subdivided as shown and the parts lettered. Any of the parts may be now selected as the **middle** part, the two parts next to it become the **adjacent** parts and the remaining two parts

4.3 SOLUTION OF THE GENERAL SPHERICAL TRIANGLE

are the **opposite** parts. Napier's rule of circular parts states that

$$\text{sIne of the mIddle part} = \text{the product of the cOsine of the Opposite parts} \quad (4.12)$$
$$= \text{the product of the tAngent of the Adjacent parts} \quad (4.13)$$

Let c be the middle part. Then

$$\sin c = \cos(90° - b)\cos(90° - C) = \sin b \sin C$$
$$= \tan(90° - A)\tan a = \cot A \tan a$$

Similarly, if $(90° - A)$ be the middle part, then

$$\sin(90° - A) = \cos a \cos(90° - C)$$
$$\cos A = \cos a \sin C$$
$$= \tan(90° - b)\tan c = \cot b \tan c$$

Note that in certain cases there may be an ambiguous solution and some indication is needed to choose the appropriate value. As a general rule, a and A, b and B, c and C take the same sign and are said to be of the same affection; i.e. a and A are both less than 90° or both more than 90°.

Example 4.7 Given $c = 96°\,57'\,00''$, $a = 35°\,14'\,30''$, $C = 90°$, then

$$\cos b = \cos c/\cos a = \cos 96.9500°/\cos 35.2417°$$
$$= -0.148\,156$$

Therefore

$$b = 98.5201° = 98°\,31'\,12''$$
$$\sin A = \sin 35.2417°/\sin 96.9500° = 0.581\,298$$
$$A = 35.5419° \text{ or } (180° - 35.5419°) = 144.4581°$$

As a is less than 90°, i.e. 35.2417°, then

$$A = 35.5419° = 35°\,32'\,31'' \text{ and}$$
$$B = \sin^{-1}(\sin b \sin A/\sin a) = 85.0608° \text{ or } 94.9392°$$

As b is greater than 90° then B is greater than 90°, i.e. $94.9392° = 94°\,56'\,21$.

Check: $\tan A \tan B \cos c$ should equal unity. Using the results obtained above,

$$\tan 35.5419° \tan 85.0608° \cos 96.9500° = -1.0003$$
or $\quad \tan 35.5419° \tan 94.9392° \cos 96.9500° = 1.0003$
or $\quad \tan 144.4581° \tan 85.0608° \cos 96.9500° = 1.0003$
or $\quad \tan 144.4581° \tan 94.9392° \cos 96.9500° = -1.0003$

It can be seen that only when (a and A) and (b and B) are of the same affection are the results acceptable. In practice for most purposes some form of visual inspection might be possible except when the angles are close to 90°.

Exercises 4.1

Solution of triangles: in each of the following questions find the missing parts.

1. Given $a = 35.3528°$, $b = 62.5917°$, $C = 90°$.
   (Ans. $c = 67.9481°$, $A = 38.6293°$, $B = 73.2992°$)

2. Given $b = 62.5917°$, $A = 38.6293°$, $C = 90°$.
   (Ans. $a = 35.3528°$, $c = 67.9481°$, $B = 73.2992°$)

3. Given $A = 56.5500°$, $B = 103.5000°$, $C = 90°$.
   (Ans. $a = 55.4676°$, $b = 106.2475°$, $c = 99.1259°$)

4. Given $c = 96.9500°$, $a = 35.2333°$, $C = 90°$.
   (Ans. $b = 98.5192°$, $A = 35.5334°$, $B = 94.9391°$)

5. Given $c = 98.5833°$, $A = 34.8667°$, $C = 90°$.
   (Ans. $a = 34.4208°$, $b = 100.4237°$, $B = 95.9367°$)

6. Given $a = 38.7000°$, $A = 45.6167°$, $C = 90°$.
   (Ans. $b = 51.6365°$ or $128.3635°$; $c = 61.0286°$ or $118.9714°$; $B = 63.6686°$ or $116.3314°$)

7. Given $a = 70.2500°$, $b = 50.3333°$, $c = 40.1667°$.
   (Ans. $A = 107.5684°$, $B = 51.2362°$, $C = 40.7955°$)

8. Given $A = 63.1764°$, $B = 72.4489°$, $C = 88.9863°$.
   (Ans. $a = 61.3833°$, $b = 69.7012°$, $c = 79.5883°$)

9. Given $a = 85.1245°$, $b = 45.4896°$, $c = 76.1589°$.
   (Ans. $A = 96.8611°$, $B = 45.2827°$, $C = 75.3548°$)

4.4 Problems involving inclined planes

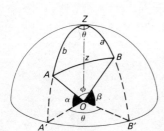

Fig. 4.6

Let the line OA represent the gradient of a plane inclined at α to the horizontal OA', whilst line OB is the gradient of a second plane inclined at β to the horizontal OB' (see Fig. 4.6). The horizontal angle between the lines OA and OB, i.e. $A'OB'$, is θ whilst the inclined lines themselves lie in a common plane and subtend an angle ϕ in the plane AOB. Assuming the point above O is Z (the zenith), the arcs $A'AZ$, $B'BZ$ and AB are all part of great circles. In the spherical triangle AZB, $Z = \theta$, $b = 90° - \alpha$, $a = 90° - \beta$, and $z = \phi$. By the cosine rule,

$$\cos \theta = \frac{\cos \phi - \cos(90° - \alpha)\cos(90° - \beta)}{\sin(90° - \alpha)\sin(90° - \beta)}$$

$$= \frac{\cos \phi - \sin \alpha \sin \beta}{\cos \alpha \cos \beta}$$

and thus

$$\cos \phi = \cos \theta \cos \alpha \cos \beta + \sin \alpha \sin \beta \qquad (4.14)$$

4.4.1 To find the maximum gradient in the plane *AOB*

In Fig. 4.7, maximum gradient OX (δ) will occur in the plane ZXX' so that $AXZ = 90°$. If α, β, θ, and ϕ are known, then

$$\sin A = \sin a \sin Z / \sin z$$
$$= \sin(90° - \beta) \sin \theta / \sin \phi$$
$$= \cos \beta \sin \theta / \sin \phi \quad (4.15)$$

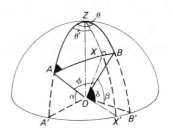

Fig. 4.7

In the right angled triangle AZX (Fig. 4.8),

$$\sin a' = \cos(90° - x) \cos(90° - A)$$
$$= \sin x \sin A$$
$$\cos \delta = \cos \alpha \sin A$$

Therefore

$$\cos \delta = \cos \alpha \cos \beta \sin \theta / \sin \phi \quad (4.16)$$

The direction of full dip is θ' from $A'O$ and thus from Fig. 4.7,

$$\sin(90° - Z') = \tan(90° - x) \tan a'$$
$$\cos Z' = \cot x \tan a'$$

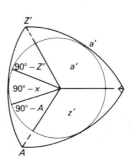

Fig. 4.8

Therefore

$$\cos \theta' = \tan \alpha \cot \delta \quad (4.17)$$

Example 4.8 A pipeline is to be laid along a bend in a mine roadway ABC. If AB falls at a gradient of 1 in 2 in a direction 036° 27′ whilst BC rises due south at 1 in 3.5, calculate the angle of bend in the pipe. (RICS/M)

Given $\alpha = \cot^{-1} 2 = 26.57°$ and $\beta = \cot^{-1} 3.5 = 15.95°$ (Fig. 4.7), by equation 4.14,

$$\phi = \cos^{-1}(\cos \theta \cos \alpha \cos \beta + \sin \alpha \sin \beta)$$
$$= \cos^{-1}(\cos 36.45° \cos 26.57° \cos 15.95°$$
$$+ \sin 26.57° \sin 15.95°)$$
$$= 35.448° = 35° 27'$$

Example 4.9 As part of a reconnaissance survey, observations were made from station A to stations B and C with a sextant and an abney level.

With the sextant $BAC = 84° 30'$
With the abney level
  Angle of depression $AB = -8° 20'$
  Angle of elevation $AC = +10° 40'$

Calculate the horizontal angle BAC which would have been measured with a theodolite at A. (RICS/M)

With the sextant,
$$\phi = BAC = 84°\,30' = 84.500°$$
With the abney level,
$$\alpha = -8°\,20' = -8.333°$$
$$\beta = 10°\,40' = 10.667°$$
By equation 4.14,
$$\theta = \cos^{-1}(\cos\phi - \sin\alpha\sin\beta)/(\cos\alpha\cos\beta)$$
$$= \cos^{-1}\left(\frac{\cos 84.500° - \sin(-8.333°)\sin 10.667°}{\cos(-8.333°)\cos 10.667°}\right)$$
$$= 82.75° = 82°\,45'$$

Exercises 4.2

1 The (non-azimuth) angle between two signals A and B was measured by a sextant from a point O, while the angular elevations of A and B above O were measured with a clinometer, with the following results.

Non azimuth angle AOB	$= 74°\,25'$
Angular elevation of A above O	$= 15°\,22'$
Angular elevation of B above O	$= 4°\,06'$

Calculate the azimuth angle AOB. (LU Ans. 74° 57′)

2 An area of ground in the form of a plane contains a path bearing 010° 00′ and rising at 8° 00′ whilst a second path bears 075° 00′ and rises at 2° 42′. Calculate the rate and direction of maximum gradient in the plane. (Ans. 8° 02′, 4° 31′)

4.5 Measurements on the earth's surface

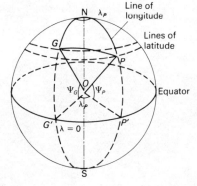

Fig. 4.9

The earth is not truly spherical but is periform although for approximate purposes or in areas of limited size, as usually applies to most engineering surveying situations, the assumption is acceptable. In terms of **geographical coordinates** a point P is defined by its **longitude**(λ) and its **latitude**(ψ). In Fig. 4.9 the meridian passing through Greenwich (G) and the North and South poles (N) and (S) cuts the equator at G' whilst the meridian through P cuts the equator at P'. The angular value $G'OP' = \lambda_p$ and represents the longitude of P east (or west) of Greenwich (the maximum value of λ is 180°). The latitude of P is represented by a plane parallel to the equator and the line of latitude is a small circle passing through P. The vertical angle $G'OG$ represents the latitude of Greenwich (ψ_G) and the angle $P'OP$ is the latitude of $P(\psi_P)$. If a great circle passes through G and P then as the lines of longitude are also great circles the area GNP is a spherical triangle.

4.5 MEASUREMENTS ON THE EARTH'S SURFACE

4.5.1 The lengths of lines on the earth's surface

Continuing only to consider the earth as a sphere there are three types of arc measurements:

1) the length of the arc of the meridian, i.e. along the arc of longitude, e.g. $A'A$;
2) the length of the arc along the parallel, i.e. along the arc of latitude, e.g. AB'';
3) the length of the arc of any great circle, e.g. AB.

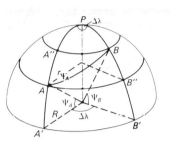

Fig. 4.10

1) Length of the arc of the meridian

In Fig. 4.10, length of arc $A'A = R\psi_A$ (4.18)

$$\text{length of arc } AP = R(90 - \psi_A) \quad (4.19)$$

$$\text{length of arc } AA'' = R(\psi_B - \psi_A) \quad (4.20)$$

$$= R \Delta\psi_{AB} \quad (4.20a)$$

Note that the angular values are expressed here as radians.

2) Length of the arc of the parallel

The arc of the parallel will have a radius

$$r_\psi = R \cos\psi \quad (4.21)$$

At the equator,

$$\text{length of arc } A'B' = R \Delta\lambda \quad (4.22)$$

At latitude ψ_A,

$$\text{length of arc } AB'' = r_{\psi A}(\lambda_A - \lambda_B) = R \cos\psi_A \Delta\lambda \quad (4.23)$$

3) The length of an arc of any great circle

Given $A(\lambda_A, \psi_A)$ and $B(\lambda_B, \psi_B)$. In the spherical triangle APB, the arc $AB = p$, the arc $AP = b = (90° - \psi_A)$ and the arc $BP = a = (90° - \psi_B)$.
Then

$$\cos P = (\cos p - \cos a \cos b)/(\sin a \sin b)$$

$$\cos p = \cos P \sin a \sin b + \cos a \cos b$$

$$= \cos\Delta\lambda \sin(90° - \psi_B)\sin(90° - \psi_A)$$

$$+ \cos(90° - \psi_B)\cos(90° - \psi_A)$$

$$= \cos\Delta\lambda \cos\psi_A \cos\psi_B + \sin\psi_A \sin\psi_B \quad (4.24)$$

The physical length of the arc is finally related to the radius of the sphere, i.e.

$$S_{AB} = R p_{\text{rad}} \quad (4.25)$$

4.5.2 Determination of azimuth

Here $\cos A = (\cos a - \cos b \cos p)/(\sin b \sin p)$

$$= (\sin\psi_B - \sin\psi_A \cos\Delta\lambda)/(\cos\psi_A \sin\Delta\lambda) \quad (4.26)$$

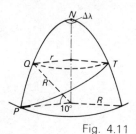

Fig. 4.11

and $\cos B = (\sin \psi_A - \sin \psi_B \cos \Delta\lambda)/(\cos \psi_B \sin \Delta\lambda)$ (4.27)

Note that
$$\phi_{AB} = A \neq \phi_{BA} = 180° - B$$

Example 4.10 A station P is located on the equator, Q and T are two stations at latitude $10°$ N, Q being due N of P and T 1062 km due E of Q (see Fig. 4.11). Assuming the earth to be a sphere of radius 6373 km, find the distance PT and its bearing.

At $\psi_Q = 10°$ N,

$r = R \cos \psi = 6373 \cos 10° = 6276.2$ km

$\Delta\lambda_{\text{rad}} = \text{arc } QT/r = 1062/6276.2 = 0.16921$

$\Delta\lambda = 9.695°$

In the spherical triangle PNT, $PN = t = 90°$, $NT = p = (90° - 10°) = 80°$ and $PNT = \Delta\lambda = 9.695°$. By the cos rule,

$n = \cos^{-1}(\cos N \cos t \cos p + \sin t \sin p)$
$\quad = \cos^{-1}(\cos 9.695° \cos 0° \cos 10° + \sin 0° \sin 10°)$
$\quad = 13.896°$

Then
$$\text{arc } PT = T n_{\text{rad}}$$
$$= 6373 \times 13.894_{\text{rad}} = 1545 \text{ km}$$
$$\phi_{PT} = \hat{P} = \sin^{-1}(\sin p \sin N/\sin n)$$
$$= \sin^{-1}(\sin 80° \sin 9.695°/\sin 13.894°)$$
$$= 43.683° = 043° 40'$$

Example 4.11 (a) A ship sails along the great circle joining two places each of latitude $45°$ N. Show that the highest latitude (L) reached during the voyage is given by
$$\cot L = \cos D/2$$
where D is the difference in longitude between the two places.
(b) Calculate the shortest distance measured along the earth's surface between New York (latitude $40° 35'$ N, longitude $74° 00'$ W) and Cape Town (latitude $33° 56'$ S, longitude $18° 26'$ E). The radius of the earth may be taken as 6370 km.
(LU)

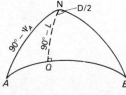

Fig. 4.12

(a) In Fig. 4.12, max $\psi = L$ occurs at Q so that $NQB = 90°$. Using Napier's mnemonics for right angled spherical triangle NQB,
$$\sin(90° - N) = \tan(90° - q) \tan b$$
(see Fig. 4.13), i.e.
$$\cos N = \cot q \tan b$$
$$\cos D/2 = \tan \psi \cot L$$
$$\cot L = \cos D/2 \quad (\text{as } \tan \psi = 1)$$

(b) By equation 4.24,

$$n = \cos^{-1}(\cos \Delta\lambda \cos \psi_A \cos \psi_B + \sin \psi_A \sin \psi_B)$$
$$= \cos^{-1}[\cos(-74°\,00' - 18°\,26')\cos 40°\,35' \cos(-33°\,56')$$
$$+ \sin 40°\,35' \sin(-33°\,56')]$$
$$= 112.949°$$

Then length $AB = 6370 \times 112.949_{\text{rad}} = 12\,557$ km.

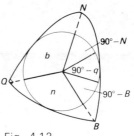
Fig. 4.13

Example 4.12 A ship sails from a place A on the equator along a great circle which makes an angle of 60° with the equator. Find the difference in longitude of A and the place B on the path of the ship where it reaches the latitude 30° N for the first time. Find also the area included by the path of the ship, the meridian through B and the equator, giving your answer to three significant figures (take $R = 6360$ km). (LU)

In the triangle ABB' (Fig. 4.14),

$$\sin b' = \tan(90° - A) \tan a'$$
$$= \cot 60° \tan 30°$$
$$b' = 19°\,28' = \Delta\lambda$$
$$\sin a' = \tan(90° - B) \tan b'$$
$$\cot B' = \sin a' \cot b' = \sin 30°\,00' \cot 19°\,28'$$
$$B' = 35°\,15'$$

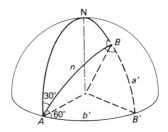
Fig. 4.14

Spherical excess (E) is given by

$$E = (A + B + B') - 180°$$
$$= (60°\,00' + 35°\,15' + 90°\,00') - 180° = 5°\,15'$$

By equation 4.2, area of triangle ABB' is given by

$$R^2(E) = 6350^2 \times 5°\,15'_{\text{rad}} = 3.7 \times 10^6 \text{ km}^2$$

Exercises 4.3

1 From the respective positions on the earth's surface of stations X and Y given below, calculate (a) the shortest distance between these two points, (b) the respective bearings of XY and YX.

	Latitude	Longitude
X	52° 36' 17" N	0° 02' 28" E
Y	43° 18' 25" N	75° 42' 56" W

Assume the radius of the earth as a sphere 6366.9 km.
(RICS/M Ans. 5465.2 km, 308° 56' 20", 68° 42' 52")

2 Two ports have the same latitude ψ and their respective longitudes differ by 2λ. Prove that the shortest route between them is $2R \sin^{-1}(\sin \lambda \cos \psi)$ where R is the radius of the earth. Find the greatest distance along the meridian between the shortest route and the parallel of latitude through the ports.

(Ans. $R[(90° - \psi) - \tan^{-1}(\cos \lambda \cot \psi)]$)

3 Two points P and Q on the earth's surface are on the same meridian, P being on the equator and Q in latitude $20°$ N. A great circle α through Q meets the equator at a point R due west of P and the angle PQR is $32°$. Calculate the shortest distance PN from P to the circle α and the angle QPN.

(LU Ans. $R \times 10°27'_{\text{rad}}$, $59° 35'$)

4.6 Convergence of the meridians (γ)

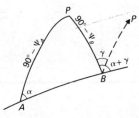

Fig. 4.15

Due to earth's curvature, the true bearing of AB and the reciprocal bearing BA do not differ by 180, i.e. $\phi_{BA} \neq \phi_{AB} \pm 180$, as the arcs of great circles change bearing continuously and thus a point on a great circle has a unique bearing. Given the geographical coordinates of $A(\psi_A, \lambda_A)$ and $B(\psi_B, \lambda_B)$, in the spherical triangle APB, $a = 90° - \psi_B$, $b = 90° - \psi_A$, and $P = \Delta\lambda$ (see Fig. 4.15). The great circle arc AB will have an azimuth $\phi_{AB} = \alpha$ at A and $(\alpha + \gamma)$ at B where γ is the angle of convergence of the meridians. Then

$$\tan\left(\frac{A+B}{2}\right) = \frac{\cos\left(\frac{a-b}{2}\right)}{\cos\left(\frac{a+b}{2}\right)} \cot\frac{P}{2} = \frac{\cos\left(\frac{\psi_A - \psi_B}{2}\right)}{\sin\left(\frac{\psi_A + \psi_B}{2}\right)} \cot\frac{\Delta\lambda}{2}$$

(see Appendix 2), but

$$\tan\left(\frac{A+B}{2}\right) = \tan[\alpha + \{180° - (\alpha + \gamma)\}]/2$$

$$= \tan(180° - \gamma)/2 = \cot \gamma/2$$

Then $\tan \gamma/2 = \dfrac{\sin\left(\dfrac{\psi_A + \psi_B}{2}\right)}{\cos\left(\dfrac{\psi_A - \psi_B}{2}\right)} \tan \dfrac{\Delta\lambda}{2}$

$$= \sin\left(\frac{\psi_A + \psi_B}{2}\right) \tan\frac{\Delta\lambda}{2} \sec\frac{\Delta\psi}{2} \qquad (4.28)$$

Where the stations are relatively close, the differences in latitude ($\Delta\psi$) and longitude ($\Delta\lambda$) will be small and the following approximation is acceptable.

$$\gamma = \sin\left(\frac{\psi_A + \psi_B}{2}\right) \Delta\lambda \qquad (4.29)$$

$$\gamma'' = \Delta\lambda'' \sin \psi_m \qquad (4.30)$$

where

$$\psi_m = \left(\frac{\psi_A + \psi_B}{2}\right) \quad \text{and}$$

$$\Delta\lambda'' = \lambda_A - \lambda_B \text{ (secs)}$$

Given the Grid coordinates of a point $A(E_A, N_A)$ and $C(E_C, N_C)$ on the central meridian at the same latitude (ψ) (Fig. 4.16) then by equation 4.30,

$$\gamma = \Delta\lambda \sin \psi$$

But $AC = r\Delta\lambda$

Then $AC = d = R \cos\psi\, \Delta\lambda$

Therefore

$$\Delta\lambda = d/(R\cos\psi) = \gamma/\sin\psi$$

Therefore

$$\gamma = d \tan\psi/R \qquad (4.31)$$
$$\gamma^\circ = 57.2958\, d \tan\psi/R \qquad (4.32)$$
$$\gamma'' = 206\,265\, d \tan\psi/R \qquad (4.32a)$$

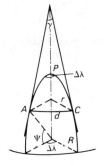

Fig. 4.16

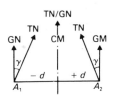

Fig. 4.17

The length d along the arc must be subjected to a local scale factor if the coordinates quoted are part of a projection system (Fig. 4.17). Related to the Ordnance Survey projection system (see p. 178),

$$d_{\text{ground}} = d_{\text{grid}}/\text{LSF}\,(F)$$

and the radius of curvature at that point may be considered to be more accurately defined as $R \simeq v$ (see p. 164). The true bearing AC = the grid bearing $AC + \gamma$. (4.33)

4.7 The spheroid

This may be defined as a mathematical figure which closely approximates to a sphere. An **ellipsoid** may be defined as the figure described by the rotation of an ellipse about one of its axes. The **geoid** (the shape of the earth) closely resembles an oblate spheroid which is similar to the rotation of an ellipse rotated about its minor axis and thus represents the best fit to the periform solid of the earth. The term geoid is defined by Cooper as the equipotential surface, arising from both the earth's attraction and rotation, which coincides with the mean level of the sea in deep water. For engineering surveying purposes, the earth's curvature is frequently negligible but where lines of, say, 10 km are involved the adoption of such an oblate spheroid is necessary and the geoid is assumed equivalent to the reference spheroid of the National Grid projection system.

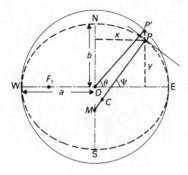

Fig. 4.18

The basic equation of the ellipse (Fig. 4.18) is given as

$$x^2/a^2 + y^2/b^2 = 1 = \sin^2\theta + \cos^2\theta \quad (4.34)$$

where a and b are the lengths of the semi-major and semi-minor axes respectively.

The eccentricity (e) of the ellipse is given as

$$e = \frac{(a^2 - b^2)^{1/2}}{a} \quad (4.35)$$

and the amount of flattening (f) is given as

$$f = (a - b)/a \quad (4.36)$$

4.7.1 Reference spheroids

	a (m)	b (m)	e	$1/f$
Airy (1849)	6 377 563	6 356 256	0.081 673	299.32
Hayford (1910)	6 378 388	6 356 912	0.081 992	297.00
Reference ellipsoid (1967)	6 378 160	6 356 775	0.081 674	298.25

Airy's spheroid was chosen by the Ordnance Survey for its transverse Mercator projection. Hayford's spheroid, known as the 'International', was recommended by the International Union of Geodesy and Geophysics in 1924, whilst in 1967 the International Association of Geodesy recommended the adoption of the Reference Spheroid.

The **spheroidal latitude** (ψ) of P is defined as the angle between the normal to the spheroid at P and the equatorial plane containing the major axis.

The **spheroidal longitude** (λ) of P is defined as the angle between the plane of the meridian through P and the Greenwich meridian plane.

4.7.2 Radii of curvature

The radius of curvature in the meridian (PC) is given as

$$\rho = a(1 - e^2)/(1 - e^2 \sin^2\psi)^{3/2} \quad (4.37)$$

The radius of curvature at right angles to the meridian (PM) is given as

$$v = a/(1 - e^2 \sin^2\psi)^{1/2} \quad (4.38)$$

then $\quad \rho = v(1 - e^2)/(1 - e^2 \sin^2\psi) \quad (4.39)$

The radius of curvature R at any latitude ψ for an azimuth ϕ is given as

$$R_\phi = \frac{\rho v}{v \cos^2\phi + \rho \sin^2\phi} \quad (4.40)$$

4.7 THE SPHEROID

The mean radius at any point is given as

$$R \simeq (\rho v)^{1/2}$$

The basic plane partial coordinates for a line of length S and bearing ϕ are

$$\Delta E = S \sin \phi \quad \text{and} \quad \Delta N = S \cos \phi$$

If $S \sin \phi$ is considered as the length of the arc along the parallel of latitude ψ then the change in longitude $(\delta \lambda)$ is given as

$$\delta \lambda'' = 206\,265\, S \sin \phi / v \cos \psi \tag{4.41}$$

or the arc length,

$$S \sin \phi = q = v \cos \psi\, \delta \lambda'' / 206\,265$$

Therefore

$$q = v \cos \psi / 206\,265 \text{ m per 1'' of longitude} \tag{4.42}$$

Similarly if $S \cos \phi$ is considered as the distance along the meridian then the change in latitude $(\delta \psi)$ is given as

$$\delta \psi'' = 206\,265\, S \cos \phi / \rho \tag{4.43}$$

or the arc length,

$$S \cos \phi = p = \rho\, \delta \psi'' / 206\,265$$

$$p = \rho / 206\,265 \text{ m per 1'' of latitude} \tag{4.44}$$

These values apply for close estimates for most lengths used in engineering surveying. For more precise computation over longer lines see Clark's formulae.

Jameson suggests the following approximate formulae (Fig. 4.19) for application on the spheroid:

$$\delta \psi'' = S \cos (\phi + \delta \phi/2)/p \tag{4.45}$$

$$\delta \lambda'' = S \sin (\phi + \delta \phi/2)/q \tag{4.46}$$

$$\delta \phi'' = \delta \lambda'' \sin \psi_m \quad \text{(see equation 4.30)} \tag{4.47}$$

where p = the length of arc for 1'' of latitude
q = the length of arc for 1'' of longitude
$\delta \phi''$ = the difference between reciprocal bearings.

Note that $\delta \phi$ is positive when $\phi > 180°$ and negative when $\phi < 180°$ (Fig. 4.20).

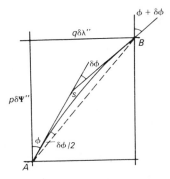

Fig. 4.19

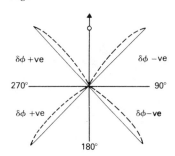

Fig. 4.20

Example 4.13 Two points A and B have the following geographical coordinates.

	ψ	λ
A	52° 21' 14.00" N	93° 48' 50.00" E
B	52° 24' 18.00" N	93° 42' 30.00" E

Given the following values,

ψ	p	q
52° 20′ 00″	30.9102 m	18.9364 m
52° 25′ 00″	30.9107 m	18.9008 m

calculate the azimuths of B from A and A from B together with the length AB.

$$\psi_m = (52°\,21'\,14'' + 52°\,24'\,18'')/2 = 52.3794°$$
$$= 52°\,22'\,46''$$

$\delta\psi_{AB} = 03'\,04'' = +184''$
$\delta\lambda_{AB} = -06'\,20'' = -380''$

By equation 4.44,

$$\delta\phi'' = \delta\lambda'' \sin\psi_m = 380 \sin 52.3794° = 301''$$

Length of arc for 1″ of latitude at ψ 52.3794° is given by

$$p = 30.9102 + (166 \times 0.0005/300) = 30.9105 \text{ m}$$

Length of arc for 1″ of longitude at ψ 52.3794° is given by

$$q = 18.9364 - (166 \times 0.0356/300) = 18.9167 \text{ m}$$

Then $(\phi + \delta\phi/2)_{AB} = \tan^{-1} \dfrac{(-380 \times 18.9167)}{(184 \times 30.9105)}$

$$= -51.6484° = \text{N}\,51°\,38'\,54.2''\,\text{W}$$
$$= 308°\,21'\,05.8''$$

As $\delta\phi = 301''$ then

$$\delta\phi/2 = 150.5'' = 02'\,30.5''$$
$$\phi_{AB} = 308°\,21'\,05.8'' + 02'\,30.5'' = 308°\,23'\,36.3''$$

and

$$\phi_{BA} = 128°\,21'\,05.8'' - 02'\,30.5'' = 128°\,18'\,35.3''$$

and $S_{AB} = 184 \times 30.9105/\cos 51.6484° = 9166.26$ m.

Example 4.14 A line AB 16 093.4 m long has an azimuth of 315° 00′ 00″. Given the latitude (ψ) of A as 54° 51′ 30.0″ N, longitude (λ) 101° 13′ 15.0″ E, find the latitude and longitude of B and the reverse azimuth assuming the following data.

ψ	p (m)	q (m)
54° 50′ N	30.9234	17.8509
54° 55′ N	30.9238	17.8141

Then $\Delta\psi = 0°05'$ $\Delta p = 0.0004$ m $\Delta q = -0.0368$ m

At 54° 51′ 30.0″,

$$p = 30.9234 + (1.5 \times 0.0004/5) = 30.9235 \text{ m}$$
$$q = 17.8509 - (1.5 \times 0.0368/5) = 17.8399 \text{ m}$$
$$\delta\psi = 16\,093.4 \times \cos 315°/30.9235 = 368'' = 06' 08.0''$$
$$\psi_m = 54°51'30'' + 03'04'' = 54°54'34''$$

At ψ_m, $p = 30.9235 + (184 \times 0.0004/300) = 30.9237$ m
$$q = 17.8399 - (184 \times 0.0368/300) = 17.8173 \text{ m}$$

then $\delta\lambda'' = 16\,093.4 \sin 315°/17.8173 = -638.69''$
$$= -10'38.69''$$
$$\delta\phi'' = 638.69 \sin 54.9094° = 522.60'' = 08'42.60''$$

The second approximation then becomes

$$(\delta\psi')'' = 16\,093.4 \cos(315°00'00'' - 04'21.3'')/30.9237$$
$$= 367.53'' = 06'07.53''$$

Then $\psi'_m = 54°51'30.0'' + 03'03.77'' = 54°54'33.77''$

At this value

$$q = 17.8141 + (26.23 \times 0.0368/300) = 17.8173 \text{ m}$$
$$(\delta\lambda')'' = 16\,093.4 \sin 314.9274°/17.8173 = -639.50''$$

Longitude $\lambda_B = 101°13'15.0'' - 10'39.5'' = 101°02'35.5''$ E
$$\delta\phi'' = 639.50 \sin 54.9094° = 523.27'' = 08'43.27''$$

Bearing $\phi_{BA} = 315°00'00'' - 180° - 08'43.27''$
$$= 134°51'16.7''$$

Latitude $\psi_B = 54°51'30.0'' + 06'07.53'' = 54°57'37.5''$ N

Exercises 4.4

1 A traverse is run from a control station A, ψ 52 20′ 00″ N and 2 00′ 00″ W, and the following partial coordinates were computed:

	ΔE (m)	ΔN (m)
AB	4310.51	4310.51
BC	5888.28	1577.77
CD	3048.00	5279.29

Given the following values,

ψ	p	q
52° 20′	30.9102	18.9364
52° 25′	30.9107	18.9008

find the change in latitude and longitude and the bearing *DC*.
(Ans. 07′ 47.14″, 09′ 49.74″, 210° 07′ 47.1″)

168 APPLICATION OF SPHERICAL TRIGONOMETRY

Given Airy's spheroid, $a = 6377\,563$ m, $b = 6356\,256$ m, $e = 0.081\,6733$, and $e^2 = 0.006\,670\,540$.

Table 4.1

ψ	v (m)	ρ (m)	$R = (\rho v)^{1/2} \rho$ (m)		q (m)
49° 00'	6389 713	6371 298	6380 499	30.8888	20.3235
50° 00'	6390 082	6372 401	6381 235	30.8942	19.9135
51° 00'	6390 449	6373 498	6381 968	30.8995	19.4974
52° 00'	6390 813	6374 587	6382 695	30.9048	19.0753
53° 00'	6391 173	6375 667	6383 415	30.9100	18.6473
54° 00'	6391 531	6376 736	6384 129	30.9152	18.2137
55° 00'	6391 884	6377 794	6384 835	30.9203	17.7743

Example 4.15 Given a line AB of length 35 245.68 m bearing 056° 27' 30.5", the geographical coordinates of A are $\psi_A = 52° 57' 19.0''$ N, $\lambda_A = 01° 09' 02.0''$ W. Based on Table 4.1, calculate the geographical coordinate of B, and the bearing BA.

By interpolation,

$$v = 6391\,173 - (41 \times 360/60)$$
$$= 6390\,927$$
$$\rho = 6375\,667 - (41 \times 1080/60)$$
$$= 6374\,929$$

By equation 4.47,

$$\delta\lambda'' = \frac{206\,265 \times 35\,245.68 \sin 56.4585°}{6390\,927 \cos 52.9553°} = 1573.8''$$
$$= 26' 13.8''$$
$$\lambda_B = -01° 09' 02.0'' + 26' 13.8'' = 00° 42' 48.2'' \text{ W}$$

By equation 4.43,

$$\delta\psi'' = \frac{206\,265 \times 35\,245.68 \cos 56.4585°}{6374\,929} = 630.1'' = 10' 30.1''$$
$$\psi_B = 52° 57' 19.0'' + 10' 30.1'' = 53° 07' 49.1'' \text{ N}$$

By equation 4.47,

$$\delta\phi'' = 1573.8 \sin 53° 02' 34.0''$$
$$= 1257.6''$$
$$\phi_{BA} = 56° 27' 30.5'' + 180° + 20' 57.6''$$
$$= 236° 48' 28.1''$$

4.8 The transverse Mercator projection

Prior to 1945 when the Davidson Report was accepted as a guide for the post war policy of the Ordnance Survey in the UK, the Cassini projection had been used. Then distances measured east-west were approximately true to scale but distances north-south became increasingly enlarged as the distance from the standard parallel increased. This had the added effect of producing errors in bearing. The transverse Mercator, modified to restrict the distortion errors, is orthomorphic, i.e. the shape is retained. The scale increases with distances from the central meridian but at any given point it is the same in all directions. It is beyond the scope of this book to deal with the mathematics of the projection system and the reader is referred to Lee and to the HMSO publication (see bibliography on p. 183), but the engineering surveyor must be conversant with the application of the projection tables; examples in the use of these follow.

4.8.1 Formulae used in calculating projection tables

In all of the tables 4.2 the lengths are based upon the international metre and reduced by the local scale factor at the central meridian F_0.

$$\text{I} = M - 100\,000$$

Note that the value of M is based upon the developed arc of the meridian from ψ_2 to ψ_1, i.e.

$$\begin{aligned}
M\psi_2 - M\psi_1 = b[&(1 + n + \tfrac{5}{4}n^2 + \tfrac{5}{4}n^3)(\psi_2 - \psi_1) \\
&- (3n + 3n^2 + \tfrac{21}{8}n^3)\sin(\psi_2 - \psi_1)\cos(\psi_2 + \psi_1) \\
&+ (\tfrac{15}{8}n^2 + \tfrac{15}{8}n^3)\sin 2(\psi_2 - \psi_1)\cos 2(\psi_2 + \psi_1) \\
&- \tfrac{35}{24}n^3 \sin 3(\psi_2 - \psi_1)\cos 3(\psi_2 + \psi_1)]
\end{aligned}$$

Here $\psi_1 = 49°$ and $\lambda_0 = -2°$.

$$\text{II} = \frac{v}{2}\sin^2 1'' \sin\psi \cos\psi \, 10^8$$

$$\text{III} = \frac{v}{24}\sin^4 1'' \sin\psi \cos^3\psi (5 - \tan^2\psi + 9\eta^2)10^{16}$$

Note that a graph based upon an equation IIIA is given but is not considered here. Also

$$\eta^2 = \frac{v}{\rho} - 1$$

Then $\quad N = (\text{I}) + P^2(\text{II}) + P^4(\text{III}) + P^6(\text{IIIA})$
where $\quad P = (\lambda_p - \lambda_0)'' 10^{-4}$.

Projection tables 4.2

$P = (\lambda_p - \lambda_0)'' 10^{-4}$ $\quad E = 400{,}000 + P(\text{IV}) - P^3(\text{V}) - P^5(\text{VI})$

$P = (\lambda_p - \lambda_0)'' 10^{-4}$ $\quad N = (\text{I}) + P^2(\text{II}) + P^4(\text{III}) + P^6(\text{IIIA})$

Latitude	(I)	Diff. 1'	(II)	Diff. 1'	(III)	Diff. 1'
52° 50	326 237.932	1853.817	3614.5265	−0.5870	0.8481	−0.0013
51	328 091.749	1853.823	3613.9385	0.5882	0.8468	0.0014
52	329 945.572	1853.827	3613.3503	0.5895	0.8454	0.0013
53	331 799.399	1853.833	3612.7608	0.5906	0.8441	0.0013
54	333 653.232	1853.839	3612.1702	0.5919	0.8428	0.0014
55	335 507.071	1853.843	3611.5783	−0.5931	0.8414	−0.0013
56	337 360.914	1853.849	3610.9852	0.5943	0.8401	0.0013
57	339 214.763	1853.854	3610.3909	0.5956	0.8388	0.0014
58	341 068.617	1853.859	3609.7953	0.5968	0.8374	0.0013
59	342 922.476	1853.864	3609.1985	0.5980	0.8361	0.0013
53° 00'	344 776.340		3608.6005		0.8348	

Latitude	(IV)	Diff. 1'	(V)	Diff. 1'	(VI)
52° 50	187 116.748	−71.628	19.7287	0.0336	0.026 95
51	187 045.120	71.645	19.7623	0.0336	0.026 94
52	186 973.475	71.660	19.7959	0.0335	0.026 92
53	186 901.815	71.676	19.8294	0.0334	0.026 90
54	186 830.139	71.692	19.8628	0.0334	0.026 89
55	186 758.447	−71.708	19.8962	0.0334	0.026 87
56	186 686.739	71.724	19.9296	0.0334	0.026 85
57	186 615.015	71.740	19.9630	0.0333	0.026 84
58	186 543.275	71.755	19.9963	0.0333	0.026 82
59	186 471.520	71.772	20.0296	0.0332	0.026 80
53° 00'	186 399.748		20.0628		0.026 79

$Q = (E - 400\,000) 10^{-6}$ $\quad \phi_p = \phi - Q^2(\text{VII}) + Q^4(\text{VIII}) - Q^6(\text{IX})$

Latitude	(VII)	Diff. 1'	(VIII)	Diff. 1'	(IX)	Diff. 1'
52° 50	3341.265	2.007	69.475	0.085	1.971	0.004
51	3343.272	2.009	69.560	0.084	1.975	0.004
52	3345.281	2.010	69.644	0.085	1.979	0.005
53	3347.291	2.012	69.729	0.085	1.984	0.004
54	3349.303	2.013	69.814	0.085	1.988	0.004
55	3351.316	2.015	69.899	0.086	1.992	0.004
56	3353.331	2.016	69.985	0.085	1.996	0.004
57	3355.347	2.018	70.070	0.086	2.000	0.004
58	3357.365	2.019	70.156	0.085	2.004	0.004
59	3359.384	2.021	70.241	0.086	2.008	0.005
53° 00'	3361.405		70.327		2.013	

$Q = (E - 400\,000) 10^{-6}$ $\quad \lambda_p = \lambda_0 + Q(\text{X}) - Q^3(\text{XI}) + Q^5(\text{XII}) - Q^7(\text{XIIA})$

Latitude	(X)	Diff. 1'	(XI)	Diff. 1'	(XII)	Diff. 1'
52° 50	53 442.5706	20.4656	978.185	1.291	33.79	0.07
51	53 463.0362	20.4859	979.476	1.294	33.86	0.08
52	53 483.5221	20.5061	980.770	1.295	33.94	0.07
53	53 504.0282	20.5264	982.065	1.298	34.01	0.08
54	53 524.5546	20.5468	983.363	1.301	34.09	0.08
55	53 545.1014	20.5672	984.664	1.303	34.17	0.07
56	53 565.6686	20.5875	985.967	1.305	34.24	0.08
57	53 586.2561	20.6079	987.272	1.307	34.32	0.08
58	53 606.8640	20.6284	988.579	1.310	34.40	0.07
59	53 627.4924	20.6488	989.889	1.312	34.47	0.08

4.8 THE TRANSVERSE MERCATOR PROJECTION

Projection tables *(contd.)*

| 53° 00' | 53 648.1412 | | 991.201 | | 34.55 | |

$P = (\lambda_p - \lambda_0)'' 10^{-4}$ $C'' = P(\text{XIII}) + P^3(\text{XIV}) + P^5(\text{XV})$

Latitude	(XIII)	Diff. 1'	(XIV)	Diff. 1'	(XV)
52° 50	7968.8152	1.7571	2.295 50	− 0.001 27	0.000 10
51	7970.5723	1.7563	2.294 23	0.001 26	
52	7972.3286	1.7557	2.292 97	0.001 27	
53	7974.0843	1.7550	2.291 70	0.001 27	
54	7975.8393	1.7543	2.290 43	0.001 28	
55	7977.5936	1.7537	2.289 15	− 0.001 27	
56	7979.3473	1.7529	2.287 88	0.001 27	
57	7981.1002	1.7523	2.286 61	0.001 27	
58	7982.8525	1.7517	2.285 34	0.001 28	
59	7984.6042	1.7509	2.284 06	0.001 27	
53° 00'	7986.3551		2.282 79		0.000 09

$Q = (E - 400\,000) 10^{-6}$ $C'' = Q(\text{XVI}) - Q^3(\text{XVII}) + Q^5(\text{XVIII})$

Latitude	(XVI)	Diff. 1'	(XVII)	Diff. 1'	(XVIII)	Diff. 1'
52° 50	42 587.397	25.702	952.124	1.306	33.715	0.076
51	42 613.099	25.722	953.430	1.308	33.791	0.075
52	42 638.821	25.742	954.738	1.311	33.866	0.076
53	42 664.563	25.762	956.049	1.312	33.942	0.077
54	42 690.325	25.781	957.361	1.315	34.019	0.076
55	42 716.106	25.801	958.676	1.318	34.095	0.077
56	42 741.907	25.821	959.994	1.320	34.172	0.077
57	42 767.728	25.841	961.314	1.322	34.249	0.077
58	42 793.569	25.861	962.636	1.324	34.326	0.077
59	42 819.430	25.881	963.960	1.327	34.403	0.078
53° 00'	42 845.311		965.287		34.481	

Latitude	(XIX)	Diff. 10'
53° 00'	0.000 426 68	− 330
10	0.000 423 38	329
20	0.000 420 09	329
30	0.000 416 80	328
40	0.000 413 52	328
50	0.000 410 24	327

$F = F_0 \left[1 + Q^2(\text{XXI}) + Q^4(\text{XXII}) \right]$

$F = F_0 \left[1 + P^2(\text{XIX}) + P^4(\text{XX}) \right]$

Latitude	(XX)	Diff. 50'
52 20	− 0.000 000 058	− 08
53 10	− 0.000 000 066	− 07

$(t_1 - T_1)'' = (2y_1 + y_2)(N_1 - N_2)(\text{XXIII}) 10^{-9}$

$(t_2 - T_2)'' = (2y_2 + y_1)(N_2 - N_1)(\text{XXIII}) 10^{-9}$

Latitude	(XXI)	Diff. 50'	(XXII)	Latitude	(XXIII)	Diff. 50'
49° 00'	0.012 2916	− 24		49° 00'	0.845 11	− 17
49 50	0.012 2892	− 24		49 50	0.844 94	− 16
50 40	0.012 2868	− 23		50 40	0.844 78	− 16
51 30	0.012 2845	− 23	0.000 0253	51 30	0.844 62	− 16
52 20	0.012 2822	− 23	Constant for the whole zone	52 20	0.844 46	− 16
53 10	0.012 2799	− 23		53 10	0.844 30	− 16
54 00	0.012 2776	− 23		54 00	0.844 14	− 15
54 50	0.012 2753	− 22		54 50	0.843 99	− 16

172 APPLICATION OF SPHERICAL TRIGONOMETRY

Also, $\text{IV} = v \sin 1'' \cos \psi \, 10^4$

$$\text{V} = \frac{v}{6} \sin^3 1'' \cos^3 \psi \left(\frac{v}{\rho} - \tan^2 \psi \right) 10^{12}$$

$$\text{VI} = \frac{v}{120} \sin^5 1'' \cos^5 \psi (5 - 18 \tan^2 \psi + \tan^4 \psi + 14\eta^2$$
$$- 58 \tan^2 \psi \eta^2 + 2 \tan^4 \psi \eta^2) 10^{20}$$

Then $E = 400\,000 + P(\text{IV}) - P^3(\text{V}) - P^5(\text{VI})$

Also $\text{VII} = \dfrac{\tan \psi \, 10^{12}}{2\rho v \sin 1''}$

$$\text{VIII} = \frac{\tan \psi}{24 \rho v^3 \sin 1''} (5 + 3 \tan^2 \psi + \eta^2$$
$$- 9 \tan^2 \psi \eta^2) 10^{24}$$

$$\text{IX} = \frac{\tan \psi}{720 \, v^5 \sin 1''} (61 + 90 \tan^2 \psi + 45 \tan^4 \psi) 10^{36}$$

Then $\psi_p = \psi' - Q^2(\text{VII}) + Q^4(\text{VIII}) - Q^6(\text{IX})$
where $Q = (E - 400\,000) \, 10^{-6}$.

Note that ψ' is obtained by interpolation from table I with N as argument.

$$\text{X} = \frac{\sec \psi \, 10^6}{v \sin 1''}$$

$$\text{XI} = \frac{\sec \psi}{6 v^3 \sin 1''} \left(\frac{v}{\rho} + 2 \tan^2 \psi \right) 10^{18}$$

$$\text{XII} = \frac{\sec \psi}{120 v^5 \sin 1''} (5 + 28 \tan^2 \psi + 24 \tan^4 \psi) 10^{30}$$

Note that a graph based upon an equation IIA is given but is not considered here.

Then $\lambda_p = \lambda_0 + Q(\text{X}) - Q^3(\text{XI}) + Q^5(\text{XII}) - Q^7(\text{XIIA})$

$\text{XIII} = \sin \psi \, 10^4$

$$\text{XIV} = \frac{\sin \psi \cos^2 \psi \sin^2 1''}{3} (1 + 3\eta^2 + 2\eta^4) 10^{12}$$

$$\text{XV} = \frac{\sin \psi \cos^4 \psi \sin^4 1''}{15} (2 - \tan^2 \psi) 10^{20}$$

Then $C'' = P(\text{XIII}) + P^3(\text{XIV}) + P^5(\text{XV})$

$$\text{XVI} = \frac{\tan \psi}{v \sin 1''} 10^6$$

$$\text{XVII} = \frac{\tan \psi}{3 v^3 \sin 1''} (1 + \tan^2 \psi - \eta^2 - 2\eta^4) 10^{18}$$

$$\text{XVIII} = \frac{\tan \psi}{15 v^5 \sin 1''} (2 + 5 \tan^2 \psi + 3 \tan^4 \psi) 10^{30}$$

Here $\quad C'' = Q(\text{XVI}) - Q^3(\text{XVII}) + Q^5(\text{XVIII})$

$$\text{XIX} \quad = \frac{\cos^2\psi \sin^2 1''}{2}(1+\eta^2)10^8$$

$$\text{XX} \quad = \frac{\cos^4\psi \sin^4 1''}{24}(5 - 4\tan^2\psi + 14\eta^2$$
$$- 28\tan^2\psi\eta^2)10^{16}$$

Then $\quad F = F_0[1 + P^2(\text{XIX}) + P^4(\text{XX})]$

$$\text{XXI} \quad = \frac{10^{12}}{2\rho v}$$

$$\text{XXII} \quad = \frac{(1+4^2)}{24\rho^2 v^2}10^{24}$$

Here $\quad F = F_0[1 + Q^2(\text{XXI}) + Q^4(\text{XXII})]$

$$\text{XXIII} = \frac{10^9}{6\rho v \sin 1''}$$

Note that 1 and 2 are the terminals of the line and a tabular function is taken out for $N_m = (N_1 + N_2)/2$ from table I. Then

$$(t_1 - T_1)'' = (2y_1 + y_2)(N_2 - N_1)(\text{XIII})10^{-9}$$

(see p. 177), and

$$(t_2 - T_2)'' = (2y_2 + y_1)(N_1 - N_2)(\text{XIII})10^{-9}$$

The tables are based upon Airy's spheroid and the following constants are used.

$a \quad = 20\,923\,713$ feet of bar $O_1 = 6377\,563.40$ m
$b \quad = 20\,853\,810$ feet of bar $O_1 = 6356\,256.91$ m
$F_0 = 0.999\,601\,2717$

Conversion factor (feet to metres) at bar $O_1 = 0.304\,800\,7491$. Also

$e^2 \quad = 0.006\,670\,540\,000\,123\,428$
$n \quad = 0.001\,673\,220\,310$

4.8.2 Examples in the use of the tables (see pp. 172–3)

1) *To compute the grid coordinates from the geographical coordinates*

The equations given are

$$E = 400\,000.000 + P(\text{IV}) - P^3(\text{V}) - P^5(\text{VI}) \quad (4.48)$$
and $\quad N = (\text{I}) + P^2(\text{II}) + P^4(\text{III}) + P^6(\text{IIIA}) \quad (4.49)$
where $\quad P = (\lambda_p - \lambda_0)'' \times 10^{-4} = P' \text{ secs} \times 10^{-4}$

Example 4.16 Given ψ_p 52° 57′ 19.0″ N, λ_p 01° 09′ 02.0″ W, then

$$P' = (-01° 09' 02.0'' + 2° 00' 00'') = 50' 58.0'' = 3058.0''$$

Therefore $P = 0.305\,80$

i.e. $\quad P = 3.058 \times 10^{-1} \quad P^2 = 9.351 \times 10^{-2}$
$\quad P^3 = 2.860 \times 10^{-2} \quad P^4 = 8.745 \times 10^{-3}$
$\quad P^5 = 2.674 \times 10^{-3} \quad P^6 = 8.178 \times 10^{-4}$

From the tables,

$$(IV) \Rightarrow 186\,615.015 - (71.740 \times 19/60) = 186\,592.297$$
$$(V) \Rightarrow 19.9630 + (0.0333 \times 0.316\dot{6}) \quad = 19.9735$$
$$(VI) \Rightarrow \quad = 0.026\,84$$

Then the Easting is, by equation 4.48,

$$E = 400\,000.000 + 57\,059.924 - 0.571 = 457\,059.353 \text{ m}$$

From the tables,

$(I) \Rightarrow 339\,214.763 + (1\,853.854 \times 0.316\dot{6}) = 339\,801.817$
$(II) \Rightarrow 3610.3909 - (0.5956 \times 0.316\dot{6}) \quad = 3610.2023$
$(III) \Rightarrow 0.8388 - (0.0014 \times 0.316\dot{6}) \quad = 0.8384$

Then the Northing is, by equation 4.49,

$$N = 339\,801.817 + 337.590 + 0.007 = 340\,139.414 \text{ m}$$

Note that these values can only be quoted as to the nearest metre as 340 139 m N, 457 059 m E.

2) *To compute the geographical coordinates from the grid coordinates*

The equations given are

$$\psi_p = \psi' - Q^2(\text{VII}) + Q^4(\text{VIII}) - Q^6(\text{IX}) \qquad (4.50)$$
and $\quad \lambda_p = \lambda_0 + Q(\text{X}) - Q^3(\text{XI}) + Q^5(\text{XII}) \qquad (4.51)$

where $Q = (E - 400\,000) \times 10^{-6}$

Example 4.17 Given the grid coordinates of P as 457 059.353 m E, 340 139.414 m N, then

$Q \quad = 0.057\,059\,353 = 5.7059 \times 10^{-2}$
$Q^2 = 3.2557 \times 10^{-3} \qquad Q^3 = 1.8577 \times 10^{-4}$
$Q^4 = 1.0600 \times 10^{-5}$
$Q^5 = 6.0481 \times 10^{-7} \qquad Q^6 = 3.4510 \times 10^{-8}$

The value of ψ' is found by interpolation for N from Table I, i.e.

$$N_p = 340\,139.414 \text{ m}$$

For ψ 52° 57' $\Rightarrow$ 339 214.763 (column (1) in Table I),

$\Delta N = 924.651$ m $\Delta\psi = 924.651/1853.854 = 0.498\,77$

$\psi' = 52° 57.498\,77' = 52° 57' 29.93''$

From the tables,

(VII) $\Rightarrow 3355.347 + (0.498\,77 \times 2.018) = 3356.354$
(VIII) $\Rightarrow 70.070 + (0.498\,77 \times 0.086) = 70.113$
(IX) $\Rightarrow 2.000 + (0.498\,77 \times 0.004) = 2.002$

Then $\psi_p = 52° 57' 29.93'' - 10.93'' - 0.00'' = 52° 57' 19.00''$

From tables

(X) $\Rightarrow 53\,586.2561 + (0.498\,77 \times 20.6079)$
    $= 53\,596.5348$
(XI) $\Rightarrow 987.272 + (0.498\,77 \times 1.307)$
    $= 987.924$
(XII) $\Rightarrow 34.32 + (0.498\,77 \times 0.08)$
    $= 34.36$

Then

$\lambda_p \Rightarrow -2° 00' 00'' + 3058.18'' - 0.18'' + 0.00''$
$= -01° 09' 02.00''$

It is possible to find the relative radii at this point ($\psi = 52° 57' 19''$) using Airy's spheroid and $e = 0.081\,673$, $e^2 = 0.006\,670\,540$. By equation 4.38,

$\nu = a/(1 - e^2 \sin^2 \psi)^{1/2}$
$= 6\,377\,548/(1 - 0.004\,249\,59)^{1/2} = 6\,391\,142$ m

By equation 4.37a,

$\rho = \nu(1 - e^2)/(1 - e^2 \sin^2 \psi)$
$= 6391\,142(1 - 0.006\,670\,540)/(1 - 0.004\,249\,59)$
$= 6375\,603$ m

Therefore

$R \simeq (\rho\nu)^{1/2} = 6383\,368$ m

4.9 Determination of approximate geographical coordinates given the grid coordinates using basic spherical trigonometry

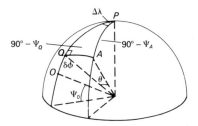

Fig. 4.21

Given a point $A(E_A, N_A)$ in Fig. 4.21, let the central meridian be OP with the origin of the projection (O) (2° W, 49° N), grid coordinates (400 km E, 100 km N), and let the line QA be perpendicular to the line OP.

The grid length is given by

$$OQ = N_A + 100 \text{ km}$$

and ground length is given by

$$OQ = \text{grid length } OQ/F_0$$

where F_0 is the scale factor at the central meridian = 0.999 601 27. The angle (in radians) subtended at the centre by this arc $\delta\psi_{\text{rad}} = \text{ground } OQ/\rho$. The latitude of Q is given by

$$\psi_Q = 49°\,00' + \delta\psi$$

and the grid length by

$$AQ = E_A - 400 \text{ km} = Y_A$$

The ground length is given by

$$AQ = Y_A/F_{AQ}$$

As the line AQ may be large, a mean scale factor should be applied (see p. 178), i.e.

$$F_{AQ} = F_0(1 + Y_A^2/6R^2) \qquad \text{by equation (4.64)}$$

The angle subtended by the arc AQ is

$$(\theta)_{\text{rad}} = \text{ground } AQ/\nu$$

In the right angled triangle QPA,

$$Q = 90° \qquad a = 90° - \psi_Q \qquad q = 90° - \psi_A$$
$$p = \theta \qquad P = \Delta\lambda$$

Then $\sin(90° - q) = \cos a \cos p$

$$\sin\psi_A = \sin\psi_Q \cos\theta \qquad (4.52)$$

also $\sin a = \cot P \tan p$

and $\tan\Delta\lambda = \tan\theta/\cos\psi_Q$

$$\lambda_A = \lambda_0 + \Delta\lambda = -2° + \Delta\lambda \qquad (4.53)$$

Example 4.18 Given A at 457 059.353 m E, 340 139.414 m N, and approximately 53° N. From Table 4.1, $\nu = 6391.173$ km, $\rho = 6375.667$ km, $R = 6383.415$ km.

$$\text{grid } OQ = 340.1394 + 100.0000 = 440.1394 \text{ km}$$
$$\text{ground } OQ = 440.1394/0.999\,601\,27 = 440.3150 \text{ km}$$
$$\delta\psi = 440.3150 \times 57.2958/6375.667 = 3.9570°$$
$$\psi_Q = 49.0000° + 3.9570° = 52.9570° = 52°\,57'\,25.1''$$
$$\text{grid } AQ = (457.0594 - 400.000) = 57.0594 \text{ km}$$

The local scale factor is given by equation 4.64:

$$F_{AQ} = 0.999\,601\,27(1 + 57.0594^2/6 \times 6383.4^2)$$
$$= 0.999\,6146$$

ground $AQ = 57.0594/0.999\,6146 = 57.0814$ km

$\theta = 57.2958 \times 57.0814/6391.173 = 0.5117° = 0°\,30'\,42''$

In the triangle QPA,

$$\psi_A = \sin^{-1}(\sin 52.9570° \cos 0.5117°) = 52.9539°$$
$$= 52°\,57'\,14.2'' \text{ N}$$
$$\delta\lambda = \tan^{-1}(\tan 0.5117°/\cos 52.9570°) = 0.8494°$$
$$= 0°\,50'\,58''$$
$$\lambda_A = -2.0000° + 0.8494° = -1.1506°$$
$$= 01°\,09'\,02'' \text{ W}$$

It can be seen from the results that the longitude value is correct when compared with the projection table value but that the latitude value is in error by 5″. This is due to the approximation of $\psi = 53°$.

4.10 Angular distortion

A straight line on the ground plots as a curve concave towards the central meridian on the transverse Mercator projection. For lines less than 10 km the error is less than 6″ but where lines are greater than this the adjustment is very necessary. In Fig. 4.22 the straight lines are grid bearings whilst the curved lines, shown broken, represent the same lines on the ground (geodesics). The theodolite at A observes the spheroidal angle YAX whilst the grid angle is BAC.

$$BAC = YAX + (t-T)_{AB} + (t-T)_{AC} \qquad (4.54)$$

where t_{AB}, t_{AC} are grid bearings and T_{AB}, T_{AC} are grid azimuths. Note that the azimuth is here defined as the spherical angle between any great circle and a meridian.

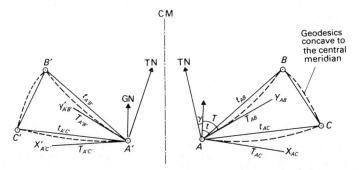

Fig. 4.22

Thus $(t-T)$ represents the difference between the two directions and is known as the 'arc to chord' correction; it is recommended that a sketch be drawn to determine whether its sign shall be positive or negative. The value of this correction is given as

$$(t-T)_{AB''} = \frac{(2Y_A + Y_B)(N_B - N_A)}{6\rho v \sin 1''} \quad (4.55)$$

$$= (2Y_A + Y_B)(N_B - N_A)(\text{XXIII}) \times 10^{-9} \quad (4.56)$$
by the projection tables

$$= (2Y_A + Y_B)(N_B - N_A) \times 845 \times 10^{-6} \quad (4.57)$$
for practical purposes

Note that the geodesic lines are always concave to the central meridian and that

$$\Sigma(t-T) = \text{spherical excess in the spherical triangle } ABC \quad (4.58)$$

4.11 Local scale factor

This is required to convert ground distance to National Grid distance and is based upon the projection system. The projection tables gives the equation for a point P as

$$F_p = F_0[1 + Q^2(\text{XXI}) + Q^4(\text{XXII})] \quad (4.59)$$

where F_0 is the local scale factor at the central meridian $= 0.999\,601\,271$ and

$$Q = Y \times 10^{-3} = (E - 400\,000) \times 10^{-6}$$

$$(\text{XXI}) = 10^{12}/2\rho v \qquad (\text{XXII}) = \frac{1 + 4\eta^2}{24\rho^2 v^2} \times 10^{24}$$

where

$$\eta^2 = \frac{v}{\rho} - 1$$

Then

$$F_p = F_0\left(1 + \frac{Q^2\,10^{12}}{2\rho v} + Q^4 \times 10^{24}\,\frac{1 + 4(v/\rho - 1)}{24\rho^2 v^2}\right) \quad (4.60)$$

For most practical work the first two terms only are required, i.e.

$$F_p = F_0(1 + Y^2/2\rho v) \simeq F_0(1 + Y^2/2R^2) \quad (4.61)$$

$$= 0.999\,6013(1 + Y^2 \times 1.228 \times 10^{-8}) \quad (4.62)$$

where $\qquad R \simeq (\rho v)^{1/2} \simeq 6381$ km

4.12 COMPUTATION OF CONVERGENCE (γ) BY PROJECTION TABLES

Example 4.19 Given $E_p = 626\,238.249$ m, latitude approximately $53°\,10'$ N.

(a) By the projection tables,

$Q = 2.262\,38 \times 10^{-1} \qquad Q^2 = 5.118\,36 \times 10^{-2}$

$Q^4 = 2.619\,76 \times 10^{-3}$

(XXI) $= 0.012\,2799$ \qquad (XXII) $= 0.000\,0253$

Then

$F_p = 0.999\,601\,271(1 + 6.285\,29 \times 10^{-4} + 6.627\,99 \times 10^{-8})$
$= 1.000\,229\,615$

(b) By equation 4.62,

$F_p = 0.999\,6013(1 + 226.238^2 \times 1.228 \times 10^{-8})$
$= 1.000\,229584$

error $= 3.1 \times 10^{-8}$

Thus for most purposes in engineering surveying the projection tables are not required to compute F_p. When the length of the line is sufficiently large (particularly E–W) then

$$F_{AB} = (F_A + 4F_C + F_B)/6 \qquad (4.63)$$

where F_A and F_B are the scale factors at the ends of the line and F_C is the scale factor at the midpoint. An alternative equation is given as

$$F_{AB} = F_0[1 + (Y_A^2 + Y_A Y_B + Y_B^2)/6R^2] \qquad (4.64)$$

Example 4.20 Given the points $A = 426\,375.36$ m E and $B = 626\,238.25$ m E then $C = 526\,306.81$ m E. By equation 4.62,

$F_A = 0.999\,6013(1 + 26.375^2 \times 1.228 \times 10^{-8})$
$= 0.999\,609\,81$

$F_C = 0.999\,6013(1 + 126.307^2 \times 1.228 \times 10^{-8})$
$= 0.999\,797\,10$

$F_B = 0.999\,6013(1 + 226.238^2 \times 1.228 \times 10^{-8})$
$= 1.000\,229\,55$

then by equation 4.63,

$F_{AB} = [0.999\,609\,81 + (4 \times 0.999\,797\,10) + 1.000\,229\,55]/6$
$= 0.999\,8380$

Then by equation 4.64,

$F_{AB} = 0.999\,6013\left(1 + \dfrac{26.375^2 + (26.375 \times 226.238) + 226.238^2}{6 \times 6381^2}\right)$

$= 0.999\,8380$

4.12 Computation of convergence (γ) by projection tables

Using geographical coordinates,
$$C'' = P(\text{XIII}) + P^3(\text{XIV}) + P^5(\text{XV})$$
where
$$P = (\lambda_p - \lambda_0) \times 10^{-4} \tag{4.65}$$

Using grid coordinates,
$$C'' = Q(\text{XVI}) - Q^3(\text{XVII}) + Q^5(\text{XVIII})$$
where
$$Q = (E - 400\,000) \times 10^{-6} \tag{4.66}$$

Given X,

$\psi_x = 52° 57' 19.0''$ $\qquad \lambda_x = 01° 09' 02.0''$ W
$E_x = 457\,059.353$ $\qquad N_x = 340\,139.414$

To find the convergence, by projection tables using ψ and λ,

$P = 3.058 \times 10^{-1}$ $P^3 = 2.860 \times 10^{-2}$ $P^5 = 2.674 \times 10^{-3}$
(XIII) $\Rightarrow 7981.1002 + (1.7523 \times 0.3166) = 7981.6551$
(XIV) $\Rightarrow 2.286\,61 - (0.001\,27 \times 0.3166) = 2.2862$

Then $\qquad C'' = 2440.79 + 0.07 = 2440.86'' = 40' 40.9''$

By projection tables using F_x,

$Q = 5.706 \times 10^{-2}$ $Q^3 = 1.858 \times 10^{-4}$ $Q^5 = 6.048 \times 10^{-7}$
(XVI) $\quad \Rightarrow 42\,767.728 + (25.841 \times 0.3166) = 42\,775.911$
(XVII) $\quad \Rightarrow 961.314 + (1.322 \times 0.3166) = 961.733$
(XVIII) $\Rightarrow 34.249 + (0.077 \times 0.3166) = 34.273$

Then $\qquad C'' = 2440.79 - 0.18 + 0.00 = 2440.61'' = 40' 40.6''$

This is probably the most accurate value.

By equation 4.30,

$$\gamma = \Delta\lambda \sin\psi$$
$\Delta\lambda = -2° 00' 00'' + 01° 09' 02.0'' = 0° 50' 58.0''$
$\qquad = 3058.0''$

$$\gamma = 3058.0 \sin 52.9553° = 2440.79'' = 40' 40.8''$$

By equation 4.32a,

$\gamma'' = 206\,265\, d \tan\psi / R$
$R = 6383.4$ km $\qquad$ (see p. 176)
$d = 57.0814$ km
$\gamma = 206\,265 \times 57.0814 \tan 52.9553 / 6383.4$
$\qquad = 2443.71'' = 40' 43.7''$

4.12 COMPUTATION OF CONVERGENCE (γ) BY PROJECTION TABLES

Example 4.21 Given the grid coordinates of two points, find the true bearing between them (Fig. 4.23).

	E (m)	N (m)
(A) College roof	457 061.180	340 246.100
(B) East Leake pillar	456 273.370	327 935.500

$\Delta E_{AB} = -787.810 \qquad \Delta N_{AB} = -12\,310.600$

grid bearing $= \tan^{-1} -787.810/-12\,310.600$
$= 183°\,39'\,41.83''$

grid length $S_{AB} = \Delta E/\sin\phi = 12\,335.781$ m
$= \Delta N/\cos\phi = 12\,335.782$ m

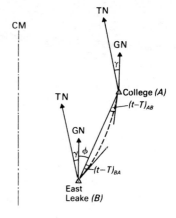

Fig. 4.23

College roof
Using projection tables, $\psi = 52°\,57'\,22.5''$

Convergence C is given by
$C'' = 2440.75'' = 0°\,40'\,40.75''$

Now, $(t-T)_{AB} = (2Y_A + Y_B)(N_B - N_A) \times 845 \times 10^{-6}$

$Y_A = 57.061\,18$ m
$2Y_A = 114.122\,36$ m $\qquad N_A = 340.246\,100$ m
$Y_B = \underline{\;56.273\,37\;}$ m $\qquad N_B = \underline{327.935\,500}$ m
$170.395\,73$ m $\qquad -12.310\,600$ m

$(t-T)_{AB} = 170.395\,73 \times -12.310\,600 \times 845 \times 10^{-6} = -1.77''$

East Leake pillar
Using projection tables,

$\psi = 52°\,50'\,44.3''$

Convergence C is given by

$C'' = 2397.69'' = 0°\,39'\,57.69''$

Now, $(t-T)_{BA} = (2Y_B + Y_A)(N_A - N_B) \times 845 \times 10^{-6}$

$Y_B = 56.273\,37$ m
$2Y_B = 112.546\,74$ m $\qquad N_B = 327.935\,500$ m
$Y_A = \underline{\;57.061\,18\;}$ m $\qquad N_A = \underline{340.246\,100}$ m
$169.607\,92$ m $\qquad 12.310\,600$ m

$(t-T)_{BA} = 169.607\,92 \times 12.310\,600 \times 845 \times 10^{-6} = 1.76''$

	Roof to pillar	Pillar to roof
Grid bearing	183° 39′ 41.83″	003° 39′ 41.83″
Convergence	+ 40′ 40.75″	+ 39′ 57.69″
$(t-T)$	− 00′ 01.77″	+ 00′ 01.76″
True bearing	184° 20′ 20.81″	004° 19′ 41.28″
	say 184° 20′ 20.8″	say 004° 19′ 41.3″

Exercises 4.5

1 The National Grid coordinates of two stations are as follows:

	E (m)	N (m)
A	326 748.21	326 402.81
B	327 435.60	326 100.45

Calculate (a) the grid bearing and length AB; (b) the scale factor applicable to the line; (c) the ground distance AB.
(Ans. 113° 44′ 35″, 750.95 m; 0.999 667; 751.20 m)

2 The difference in longitude between two stations is 0° 43′ 20″ and the latitude of each station is N 54° 00′. Calculate the convergence between the meridians through the stations.
(Ans. 35′ 03″)

3 The National Grid coordinates of a station are 626 238.25 m E, 302 646.42 m N. Calculate the approximate latitude and longitude. (Ans. ψ 52° 34′ 27″, λ 1° 20′ 16″ E)

4 From a point X in latitude 50° 00′ 00″ N a straight line XY 11 150 m long is set out with an azimuth of 60° 00′ at X. Assuming the earth to be a sphere so that 30.94 m at the surface subtends 1″ at the centre, by solving the spherical triangle PXY, compute (a) the azimuth of YX at Y; (b) the difference in longitude between X and Y; (c) the latitude of Y.

Check your computation of the azimuth by calculating the convergence using an equation applicable to short lines.
(Ans. 240° 06′ 12.6″, 0° 08′ 06″, 50° 03′ 00″)

5 A traverse starts from station A which is known to be in latitude 53° 46′ 22″ N and the first leg has a measured length of 11 367.2 m. The azimuth of B from A is 62° 30′ 30″. The second leg BC of the traverse deflects right 24° 27′ 36″. Determine the forward azimuth of BC and the approximate latitude of station B given the following linear values of 1″ of arc measured along the meridian and parallel.

Latitude	1″ of latitude (m)	1″ of longitude (m)
53° 45′	30.9177	18.3256
53° 50′	30.9182	18.2893

(LU 87° 05′ 30″, 53° 49′ 12″)

6 Two points A and B have the following coordinates.

	Latitude	Longitude
A	52° 21′ 14″ N	93° 48′ 50″ E
B	52° 24′ 18″ N	93° 42′ 30″ E

Given the following values

Latitude	1″ of latitude	1″ of longitude
52° 20′	30.423 45 m	18.638 16 m
52° 25′	30.423 87 m	18.603 12 m

find the azimuths of B from A and A from B also the distance AB. (LU Ans. 308° 23′ 36″, 128° 18′ 35″, 9021.9 m)

7 The following data refers to two triangulation points.

	E (m)	N (m)	Latitude
A	457 061.18	340 246.10	52° 57′ 22.5″ N
B	456 273.37	327 935.50	52° 50′ 44.3″ N

Taking the radius of the earth as 6384.1 km, calculate (a) the grid bearing AB; (b) the approximate convergence of the National Grid and true meridians at A; (c) the $(t-T)$ correction at A; (d) the 'true bearing' of the line AB.
(RICS/M Ans. 183° 39′ 41.8″, 40′ 42.7″, −1.8″, 184° 20′ 22.7″)

8 The geographical coordinates of two stations X and Y are as follows.

	Latitude (ψ)	Longitude (λ)
X	52° 06′ 35.3″ N	02° 06′ 13.8″ W
Y	52° 10′ 05.9″ N	02° 03′ 17.4″ W

Compute the distance XY on the assumption that the earth is a sphere of radius 6367.272 km   (RICS/M 7285.3 m)

Bibliography

ALLAN, A. L., HOLLWEY, J. R., and MAYNES, J. H. B., *Practical Field Surveying and Computations*. Heinemann (1973)
CLARK, D., (revised by Jackson, J. E.), *Plane Geodetic Surveying for Engineers*, 2
CLARK, D., *Plane and Geodetic Surveying*, **II**. Constable and Co. (1963)
COOPER, M. A. R., *Fundamentals of Survey Measurement and Analysis*. Crosby Lockwood Staples (1974)
JAMESON, A. H., *Advanced Surveying*. Pitman (second edition 1948)
LEE, L. P., The transverse Mercator projection of the spheroid. *Empire Survey Review*, 58 (1945)
MALING, D. H., *Coordinate Systems and Map Projections*. George Philip and Son (1973)
HMSO, *Constants, Formulae and Methods used in the Transverse Mercator Projection* (1950)

5
Circular curves

5.1 Definitions

A circular curve may be defined by one of the following

1) the radius (R);
2) the degree of the curve (D);
3) the equations of the curve in cartesian coordinate form.

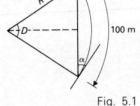

Fig. 5.1

The radius (R) is often preferred as the initial design starts by plotting the centre line route to a horizontal scale of 1 in H (say 1/2500) using railway curves. These are defined by the number (n) equivalent to the true radius of the arc in mm. Then

$$R = nH/1000 \text{ metres} \tag{5.1}$$

The degree of the curve (D), in metric terms, is the angle subtended at the centre of the curve by a 100 m arc (Fig. 5.1). Then

the arc $= RD_{\text{rad}} = 100 \text{ m}$

$$D_{\text{rad}} = 100/R \tag{5.2}$$

$$D° = 100 \times 180/(R\pi) = 5729.578/R \tag{5.3}$$

Therefore $\quad R = 5729.578/D \text{ metres} \tag{5.4}$

Example 5.1

D (degree)	R (m)
1	5729.578
2	2864.789
10	572.958
28.6479	200.000

Note that the tangential angle $\alpha = D/2$. \quad (5.5)

The equation of the curve is given as

$$(x - x_0)^2 + (y - y_0)^2 = R^2 \tag{5.6}$$

where the coordinates of the centre O are given as x_0, y_0 (Fig. 5.2).

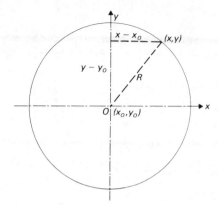

Fig. 5.2

Then $x^2 - 2x_0 x + x_0^2 + y^2 - 2y_0 y + y_0^2 - R^2 = 0$
i.e. $x^2 + y^2 - 2x_0 x - 2y_0 y + (x_0^2 + y_0^2 - R^2) = 0$ (5.7)

This is generally given as

$$x^2 + y^2 + 2gx + 2fy + c = 0 \qquad (5.8)$$

where $g = -x_0, f = -y_0, c = x_0^2 + y_0^2 - R^2$, and

$$x^2 + y^2 - R^2 = 0 \qquad \text{when } x_0 = y_0 = 0 \qquad (5.9)$$

Note that the comparable equation of a straight line joining two given points is

$$(y - y_1)/(x - x_1) = (y_2 - y_1)/(x_2 - x_1) = m \qquad (5.10)$$

i.e. $y - y_1 = m(x - x_1)$
then $y = m(x - x_1) + y_1 = mx - mx_1 + y_1$
      $= mx + c$
where $c = y_1 - mx_1$ (5.11)

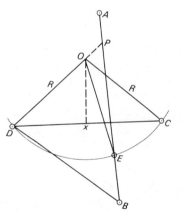

Fig. 5.3

Example 5.2 A railway boundary CD, in the form of a circular arc, is shown in Fig. 5.3 intersected by a farm boundary BA at E. Calculate the coordinates of E given the coordinates of A, B, C and D and the radius of the curve as 200 m.

	E (m)	N (m)
A	3010.28	8780.68
B	3125.36	8065.32
C	3203.72	8471.84
D	2840.48	8429.72

From the coordinates given, the bearing of DC is

$$\phi_{dc} = 83.386°$$

The length of DC is

$$S_{dc} = 365.674 \text{ m} \qquad S_{dc}/2 = 182.837 \text{ m}$$
$$\phi_{ab} = 170.861°$$

and in triangle DOX (with O as centre),

$$XO = (R^2 - 182.837^2)^{1/2} = 81.059 \text{ m}$$
$$\text{angle } D = \sin^{-1}(81.059/R) = 23.910°$$
thus $\quad \phi_{do} = \phi_{dc} - D \qquad\qquad = 59.476°$

To obtain the coordinates of O, for line DO

$$S = 200.00 \text{ m} \quad \text{and} \quad \phi = 59.476°$$
then $\quad \Delta E = 172.28 \quad$ giving $E_o = 3012.76 \text{ m}$
$\quad\quad\; \Delta N = 101.58 \quad$ giving $N_o = 8531.30 \text{ m}$
$\quad\quad\; \phi_{db} = 141.982° \qquad S_{db} = 462.541 \text{ m}$

In triangle DPB,

angle $D = 141.982 - 59.476 = 82.506°$
angle $B = 350.861 - 321.982 = 28.879°$
$DP = DB \sin B / \sin(D + B) = 239.907 \text{ m}$

To obtain the coordinates of P, for line DP,

$$S = 239.907 \text{ m} \qquad \phi = 59.476°$$
$$\Delta E = 206.66 \quad \text{giving } E_p = 3047.14 \text{ m}$$
$$\Delta N = 121.85 \quad \text{giving } N_p = 8551.57 \text{ m}$$

In triangle POE,

$$\phi_{pe} = \phi_{ab} = 170.861° \qquad S_{pb} = 492.501 \text{ m}$$
$$PO = 239.907 - 200.000 = 39.907 \text{ m}$$
$$OE = R = 200.00 \text{ m}$$
angle $P = \phi_{pd} - \phi_{pb}$
$\qquad\quad = 239.476 - 170.861 = 68.615°$
angle $E = \sin^{-1}(OP \sin P / 200) = 10.708°$
angle $O = 180° - (E + P) = 100.677°$
thus $\quad \phi_{oe} = \phi_{op} + O = 59.476 + 100.677 = 160.153°$

To obtain the coordinates of E, for line OE

$$\Delta E = 67.901 \qquad \text{giving } E_e = 3080.66 \text{ m}$$
$$\Delta N = -188.121 \quad \text{giving } N_e = 8343.18 \text{ m}$$

Alternative solution
Using the equation of the circle (5.9),

$$x^2 + y^2 - R^2 = 0 \qquad \text{centre } (0, 0)$$

and the straight line (equation 5.11),

$$y = mx + c$$

By substitution,

$$x^2 + (mx + c)^2 - R^2 = 0$$
$$x^2(1 + m^2) + x(2mc) + (c^2 - R^2) = 0$$

Here $m = (y_2 - y_1)/(x_2 - x_1)$
$= (N_b - N_a)/(E_b - E_a) = \cot \phi_{ab} = -6.2162$

and $c = y - mx$
$= (N_a - N_o) - m(E_a - E_o) = N_a - mE_a = 233.9638$

Then $1 + m^2 = 39.6411$ ⎫
$2mc = -2908.7315$ ⎬ the coefficients of the quadratic equation in x
$c^2 - R^2 = 14739.0597$ ⎭

Solving,

$$39.6411 x^2 - 2908.7315 x + 14739.0597 = 0$$

gives $x = 67.90$ or 5.48
and $y = -188.12$

As the coordinates of O are 3012.76 m E, 8531.30 m N (as computed previously), the coordinates of E are

$$3012.76 + 67.90 = 3080.66 \text{ m E}$$
$$8531.30 - 188.12 = 8343.18 \text{ m N}$$

5.2 Through chainage

This represents the length of road or rail from some terminus. In the past, the term chainage related to the active use of the chain as a unit of measurement and as a measuring device. This term is universally understood and there is little reason to change it with the advent of use of the metre in surveying. Pegs are placed at 'stations' and any point on the construction can be defined by reference to these 'stations'. Whilst the basic metric station unit is 100 m, this is too large for practical purposes and sub-multiples are used, e.g. 5, 10, 20 or 25 m (Fig. 5.4). The old format

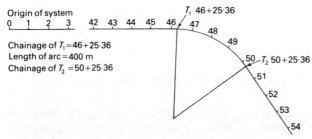

Fig. 5.4 Through chainage

is often retained, e.g. $46 + 25.36 \text{ m} \equiv 4625.36 \text{ m}$. If the chainage of the tangent point $T_1(C_{T1}) = 4625.36 \text{ m}$
and the arc $T_1T_2 = \underline{400.00 \text{ m}}$
then the chainage of T_2 $(C_{T2}) = 5025.36 \text{ m}$

Note that if the standard chord is 20 m then the first peg on the curve will be 4640.00 m.

5.3 Length of arc (L)

$$L = 2\pi R \times \theta/360 \qquad (5.12)$$
or $\quad L = R\theta_{rad} = R\theta°/57.2958 \qquad (5.13)$

Example 5.3 If $\theta = 30° 26'$ and $R = 100 \text{ m}$ then
$$R = 2 \times 3.142 \times 100 \times 30.433/360 = 53.12 \text{ m}$$
or $\quad R = 100 \times 30.433/57.2958 \qquad = 53.12 \text{ m}$

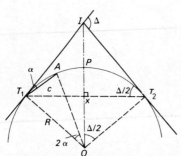

Fig. 5.5

5.4 Geometry of the curve

In Fig. 5.5, T_1 and T_2 are tangent points, I is the intersection point, Δ is the deflection angle at I (deflection may be left or right), and

$$IT_1 = IT_2 = \text{the tangent length} = R \tan \Delta/2 \qquad (5.14)$$
$$T_1T_2 \text{ is the long chord} = 2TX = 2R \sin \Delta/2 \qquad (5.15)$$
$$T_1A = \text{the chord } c = 2R \sin \alpha \qquad (5.16)$$

(If $c \leqslant R/20$ (see Section 5.5) α will be small and the arc approximates to a chord, i.e. $\sin \alpha \simeq \alpha_{rad}$.)

$$c = 2R \alpha_{rad}$$

The deflection angle is
$$\alpha = \sin^{-1}(c/2R) \qquad (5.17)$$
$$\alpha'' \simeq 206\,265\, c/2R \qquad (5.18)$$
$$\alpha° \simeq 57.2958\, c/2R \qquad (5.19)$$
$$IO = R \sec(\Delta/2) \qquad (5.20)$$
$$IP = IO - PO = R \sec(\Delta/2) - R = R(\sec \Delta/2 - 1) \qquad (5.21)$$
$$PX = PO - XO = R - R \cos(\Delta/2) = R(1 - \cos \Delta/2) \qquad (5.22)$$
$$= R \text{ versine}(\Delta/2) \qquad (5.23)$$

5.5 The arc/chord approximation

For through chainage purposes the standard chord (c_s) is chosen as a sub-multiple of 100 m and, in order that accuracy is not lost, a chord/radius (c/R) ratio $\leqslant 1/20$ is frequently chosen. The

5.5 THE ARC/CHORD APPROXIMATION

length of the chord is $c = 2R \sin \alpha$ and the length of the arc $A = 2R\alpha_{rad}$. Writing

$$\sin \alpha = \alpha - \alpha^3/3! + \alpha^5/5! - \alpha^7/7! \quad \text{etc.}$$

(see Appendix 3) then

$$\begin{aligned} c = 2R \sin \alpha &= 2R(\alpha - \alpha^3/3! + \alpha^5/5! + \ldots) \\ &= 2R(\alpha - \alpha^3/6 + \alpha^5/120 + \ldots) \\ &= A - A^3/24R^2 + A^5/1920R^4 + \ldots \end{aligned}$$

Assuming an accuracy of 1/10 000, then

$$c/A = 1/1.0001 = 0.9999$$

i.e.
$$c/A \simeq 2R(\alpha - \alpha^3/6)/2R\alpha = 0.9999$$
$$1 - \alpha^2/6 = 0.9999$$
$$\alpha^2 = 0.0006$$
$$\alpha_{rad} = 0.0245$$
$$\alpha° = 57.2958\, \alpha_{rad} = 1.4037° = 1°\, 24'\, 13''$$

But $\quad c/2R \simeq \alpha_{rad} = 0.0245 = 1/40.82$
Therefore

$$c/R = 1/20.4$$

Thus for an accuracy of

	1/10 000	$c/R \leqslant 1/20$
and for	1/5000	$c/R \leqslant 1/14$
and for	1/2500	$c/R \leqslant 1/10$

Example 5.4 Given the chainage of the intersection point (C_I) = 2637.462 m, the radius (R) = 200 m, and deflection angle (Δ) = 30° 27′ 40″, then

$$\begin{aligned} \text{tangent length} &= R \tan(\Delta/2) \\ &= 200 \tan 15.231° = 54.453 \text{ m} \\ \text{arc length } T_1T_2 &= R\, \Delta_{rad} \\ &= 200 \times 0.532 = 106.329 \text{ m} \\ \text{chainages}\quad I &= 2637.462 \text{ m} \\ \text{tangent length } TL &= -54.453 \text{ m} \\ T_1 &= 2583.009 \text{ m} \\ \text{arc length } T_1T_2 &= 106.329 \text{ m} \\ T_2 &= 2689.338 \text{ m} \end{aligned}$$

If c_s = standard chord = (say) 200/20 = 10 m then chainage of P_1 will be 2590.00.

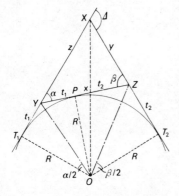

Fig. 5.6

5.6 Special problems

5.6.1 To pass a curve tangential to three given straights

With reference to Fig. 5.6,

$$R = [(x+y+z)/2] \cot(\Delta/2) \quad \text{or} \quad s\cot(\Delta/2) \quad (5.24)$$
$$R = (\text{Area of } \Delta\, XYZ)/(s-x) \quad (5.25)$$
$$R = (YZ)/(\tan\alpha/2 + \tan\beta/2) \quad (5.26)$$

Example 5.5 The coordinates of three stations A, B and C are as follows.

	E (m)	N (m)
A	1263.13	1573.12
B	923.47	587.45
C	1639.28	722.87

The lines AB and AC are to be produced and a curve set out so that the curve will be tangential to AB, BC and AC. Calculate the radius of the curve.

Bearing AB is

$$\phi_{ab} = \tan^{-1}(923.47 - 1263.13)/(587.45 - 1573.12)$$
$$= 199.014°$$

Length AB is

$$S_{ab} = (339.66^2 + 985.67^2)^{1/2} = 1042.55 \text{ m}$$

Similarly, $\phi_{ac} = 156.135°$ $\quad \phi_{bc} = 79.287°$
$\quad S_{ac} = 929.74$ m $\quad S_{bc} = 728.51$ m

Then the deflection angle is

$$\Delta = 156.135° - 19.014° = 137.121° = 137°\,07'\,15''$$

To find the radius R, by equation 5.24

$$R = s \cot(\Delta/2)$$
$$= [(1042.55 + 728.51 + 929.74)/2] \cot 68.561°$$
$$= 1350.40 \cot 68.561° = 530.28 \text{ m}$$

and by equation 5.25,

$$R = \frac{(bc \sin \Delta)/2}{s - x}$$

$$= \frac{929.74 \times 1042.55 \times 0.680\,452/2}{1350.40 - 728.51} = 530.28 \text{ m}$$

and by equation 5.26,

$$R = BC/(\tan \alpha/2 + \tan \beta/2)$$
$$\alpha = 79.287° - 19.014° = 60.273° \quad \alpha/2 = 30.137°$$
$$\beta = 336.135° - 259.287° = 76.848° \quad \beta/2 = 38.424°$$

Then $R = 728.51/(\tan 30.137° + \tan 38.424°)$
$$= 530.28 \text{ m}$$

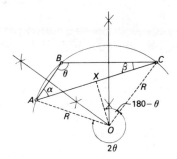

Fig. 5.7

5.6.2 To pass a curve through three points

With reference to Fig. 5.7, by the sine rule

$$a/\sin A = b/\sin B = c/\sin C = 2R$$

i.e. here

$$a/\sin \alpha = b/\sin \theta = c/\sin \beta = 2R$$

Example 5.6 The coordinates of two points B and C relative to A are

B 449.95 m E   536.23 m N
C 1336.28 m E   692.34 m N

Calculate the radius of the circular curve passing through the three points.

The bearing AB is

$$\phi_{ab} = \tan^{-1} 449.95/536.23 = 040.000°$$
and $\quad \phi_{bc} = \tan^{-1} 886.33/156.11 = 080.011°$
Then $\quad \theta = \phi_{ba} - \phi_{bc} = 220.000° - 080.011° = 139.989°$

Length AC is

$$S_{ac} = (692.34^2 + 1336.28^2)^{1/2} = 1504.98 \text{ m}$$
Then $\quad R = 1504.98/(2 \sin 139.989)$
$$= 1170.40 \text{ m}$$

Exercises 5.1

1 It is required to range a simple curve which will be tangential to three straight lines YX, PQ, and XZ, where PQ is a straight line joining the two intersecting lines YX and XZ. Angles $QPY = 134° 50'$, $ZXY = 72° 30'$, $ZQP = 117° 40'$, and the distance $XP = 57.50$ m. Compute the tangent distance from X along the straight YX and the radius of curvature.

(ICE Ans. 82.73 m, 60.66 m)

2 A circular road has to be laid out so that it shall be tangential to each of three lines DA, AB, and BC. Given the coordinates and bearings as follows, calculate the radius of the circle.

	E (m)	N (m)
A	871.24	970.66
B	941.61	822.03

Bearing $DA = 114°\,58'\,10''$
Bearing $CB = 054°\,24'\,10''$ (Ans. 137.48 m)

3 Three points A, B and C lie on the centre line of an existing mine roadway. A theodolite is set at B and the following observations were taken on to a vertical staff.

Staff at	Horizontal circle	Vertical circle	Stadia	Staff readings Collimation
A	002° 10' 20''	+2° 10' 00''	2.082/1.350	1.716
C	135° 24' 40''	−1° 24' 00''	2.274/1.256	1.765

If the multiplying constant is 100 and the additive constant zero, calculate (a) the radius of the circular curve which will pass through A, B and C; (b) the gradient of the track laid from A to C, if the instrument height is 1.573 m.

(RICS/M Ans. 110.4 m; 1 in 34)

5.6.3 To pass a curve through a given point

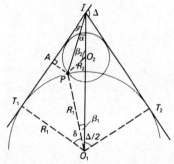

Fig. 5.8

(a) Given the coordinates of P relative to I, i.e. IA and AP, (Fig. 5.8)

$$\theta = \tan^{-1}(AP/IA)$$
$$IP = (AP^2 + IA^2)^{1/2}$$

(b) Given the polar coordinates of P from the intersection point I (IP, θ),

$$\alpha = 90° - (\Delta/2 + \theta)$$
$$OI = R \sec \Delta/2$$
$$OI/OP = (R \sec \Delta/2)/R = \sin(\alpha + \beta)/\sin \alpha$$

and $\sin(\alpha + \beta) = \sin \alpha \sec \Delta/2$
$$= \sin[180° - (\alpha + \beta)] \quad (5.27)$$

Thus there are two values of β

$$R/\sin \alpha = IP/\sin \beta$$
$$R = IP \sin \alpha \operatorname{cosec} \beta \quad \text{(2 values)} \quad (5.28)$$

If $\alpha = 0$ then P lies on the line IO and

$$OI = R \sec \Delta/2 = IP \pm R$$
$$R = IP/(\sec \Delta/2 \pm 1) \quad \text{(2 values)} \quad (5.29)$$

If $\theta = 90°$ then

$$T_1 O = R$$
$$QP = T_1 I = R \sin \delta = R \tan \phi/2$$

So $\sin \delta = \tan \phi/2$

$$T_1 Q = IP = R - R \cos \delta = R(1 - \cos \delta)$$

and $R = IP/\text{vers } \delta \quad \text{(2 values)} \quad (5.30)$

Example 5.7 A circular railway curve has to be set out to pass through a point which is 40.0 m from the intersection point and equidistant from the tangent points. The straights are deflected through 46° 40'. Calculate the radius of the curve and the tangent length.

Taking the maximum value only, P lies on the line IO and by equation 5.29,

$$R = IP/(\sec \Delta/2 \pm 1)$$
$$= 40.0/(\sec 23.33 \pm 1) = 19.148 \text{ or } 449.094$$
$$\text{say } 449.1 \text{ m}$$
$$T_1 I = R \tan \Delta/2 = 193.7 \text{ m}$$

Example 5.8 The bearings of two lines AB and BC are 036° 36' and 080° 00' respectively. At a distance of 276.00 m from B towards A and 88.00 m at right angles to the line AB, a station P has been located. Find the radius of the curve to pass through the point P and also touch the two lines. (RICS/M)

With reference to Fig. 5.9, the polar coordinates are

$$\theta = \tan^{-1} 88/276 = 17.684°$$
$$BP = (88^2 + 276^2)^{1/2} = 289.69 \text{ m}$$
$$\alpha = 90° - (\Delta/2 + \theta) = 90° - [(80.00 - 36.60)/2 + 17.68]$$
$$= 50.62°$$

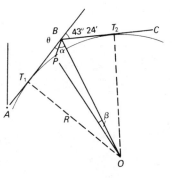

Fig. 5.9

By equation 5.27,

$$\alpha + \beta = \sin^{-1} (\sin \alpha/\cos \Delta/2)$$
$$= \sin^{-1} (\sin 50.62/\cos 21.70) = 56.30°$$
$$\beta = 56.30° - 50.62° = 5.68°$$
$$R = IP \sin \alpha/\sin \beta$$
$$= 289.69 \sin 50.62°/\sin 5.68° = 2262.4 \text{ m}$$

Exercises 5.2

1 In setting out a circular curve it is found that the curve must pass through a point 15.24 m from the intersection point and equidistant from the tangents. The chainage of the intersection point is 85 + 58.78 m and the deflection angle is 28° 00′. Calculate the radius of the curve, the chainage at the beginning and end of the curve and the degree of curvature based upon 100 m arc.   (ICE 497.82 m; 84 + 34.66 m, 86 + 77.94 m; 11° 30′)

2 Two straights, intersecting at a point B, have the bearings BA 270° 00′, BC 110° 00′. They are to be joined by a circular curve, but the curve must pass through a point D which is 45.72 m from B and the bearing BD is 260° 00′. Find the required radius, the tangent distances, the length of the curve and the deflection angle for a 20 m chord.
(LU 950.74, 167.64, 331.87, 0° 36′ 09″)

3 The coordinates of the intersection point I of two railway straights, AI and IB, are 1000.00 m E, 1000.00 m N. The bearing of AI is 90° 00′ and that of IB is 57° 14′. If a circular curve is to connect these straights and if this curve must pass through 696.90 m E, 1020.40 m N, find the radius of the curve.

Calculate, also, the coordinates of the tangent point on AI and the deflection angles necessary for setting out 100 m chords from this tangent point. What would be the deflection angle to the other tangent point and what would be the final chord length?
(LU 2000.25 m, 411.94 m E, 1000.00 m N, 1° 25′ 56″, 16° 22′ 59″
43.91 m)

5.7 Location of the curve

If no part of the curve or the straights exist, setting out is related to a development plan, controlled by traverse and/or topographical detail. Frequently the curves are defined by a series of intersection points (i.e. their coordinates have been estimated from the development plan) and thus the bearings of the straights may be computed. The radii are usually subjected to topographical constraints and the designer of the route may use railway curves (see p. 184). Subsequently all of the relevant coordinates, i.e. tangent lengths and points on the curve, may be computed and thereafter related to the control points.

Example 5.9 The data below Fig. 5.10 relating the intersection points to the control points are given (see Fig. 5.10).

Using railway curve No. 160 on a plan 1/2500, calculate the coordinates of T_1 and T_2 to the drawn curve which is tangential to I_1I_2 and I_2I_3 and the setting out bearings at A, B, C and D to locate these points.

5.7 LOCATION OF THE CURVE 195

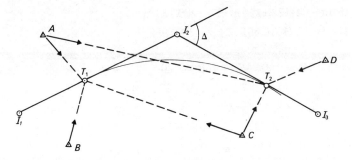

Fig. 5.10

		E (m)	N (m)
Intersection points	I_1	1000.000	1000.000
	I_2	1452.500	1164.700
	I_3	1880.300	997.200
Control points	A	1280.126	1200.134
	B	1242.117	950.123
	C	1521.463	1001.148
	D	1824.987	1150.954

The bearing I_1/I_2 is

$$\phi_{I1/I2} = \tan^{-1}(452.500/164.700) = 70.000°$$
$$\phi_{I2/I3} = \tan^{-1}(427.800/-167.500) = 111.382°$$
$$\Delta = 41.382°$$

By equation 5.1,

$$\text{radius of the curve} = nH/1000 = 160 \times 2500/1000$$
$$= 400.000 \text{ m}$$
$$\text{tangent length } T_1 I_2 = R \tan \Delta/2 = 400 \tan 20.691°$$
$$= 151.076 \text{ m}$$

To compute the coordinates of T_1, for line $I_2 T_1$

$S = 151.076$ m $\quad \phi = 250.000°$
$\Delta E = -141.965$ m $\quad E_{T1} = 1310.535$ m
$\Delta N = -51.671$ m $\quad N_{T1} = 1113.029$ m

To compute the coordinates of T_2, for line $I_2 T_1$,

$S = 151.076$ m $\quad \phi = 111.382°$
$\Delta E = -140.677$ m $\quad E_{T2} = 1593.177$ m
$\Delta N = -55.080$ m $\quad N_{T2} = 1109.620$ m

The relevant lengths and bearings are then computed for setting out from the control points.

From A (1280.126 m E,   1200.134 m N)
To T_1 (1310.535 m E, 1113.029 m N)
       $S = 92.260$ m
       $\phi = 160.756° = 160° 45' 20''$
To T_2 (1593.177 m E, 1109.620 m N)
       $S = 325.874$ m
       $\phi = 106.126° = 106° 07' 35''$
From B (1242.117 m E,   950.123 m N)
To T_1
       $S = 176.690$ m
       $\phi = 022.782° = 022° 46' 54''$
From C (1521.463 m E,   1001.148 m N)
To T_1
       $S = 238.763$ m
       $\phi = 297.942° = 297° 56' 33''$
To T_2
       $S = 130.035$ m
       $\phi = 033.470° = 033° 28' 11''$
From D (1824.987 m E,   1150.954 m N)
To T_2
       $S = 235.466$ m
       $\phi = 259.890° = 259° 53' 23''$

T_1 may be located by cutting in from A and B and then checked from C. T_2 may be located by cutting in from C and D and then checked from A.

If the straights exist
1) Locate the intersection point.
2) Measure the deflection angle Δ.
3) Compute the tangent length (the radius must be known).
4) Set out the tangent lengths T_1 and T_2 from I.

If the intersection point is inaccessible
Select two stations on each straight, e.g. A, B, C and D in Fig. 5.11. Then carry out the following procedure.

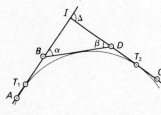

Fig. 5.11

1) Measure or compute α, β and the length BD.

   Solve the triangle BID:

   $BI = BD \sin\beta/\sin\Delta = K \sin\beta$
   $DI = BD \sin\alpha/\sin\Delta = K \sin\alpha$
   $T_1 I = T_2 I = R \tan\Delta/2$
   $T_1 B = T_1 I - BI$
   $T_2 D = T_2 I - DI$

2) If the coordinates of A, B, C and D are known then, using equation 2.3 to obtain I,

$$\Delta N_{ai} = (\Delta E_{ac} - \Delta N_{ac} \tan \phi_{cd})/(\tan \phi_{ab} - \tan \phi_{cd})$$

but $\tan \phi_{ab} = \Delta E_{ab}/\Delta N_{ab} = \text{say } m$

and $\tan \phi_{cd} = \Delta E_{cd}/\Delta N_{cd} = \text{say } n$

Then $\Delta N_{ai} = (\Delta E_{ac} - n \Delta N_{ac})/(m - n)$ (5.31)

and $\Delta E_{ai} = \Delta N_{ai} \tan \phi_{ab}$ (5.32)

$= m \Delta N_{ai}$

The coordinates of I are then given as

$$E_i = E_a + \Delta E_{ai} \quad \text{and} \quad N_i = N_a + \Delta N_{ai}$$

Example 5.10 Stations A, B, C and D lie on existing straights and have the following coordinates.

	E (m)	N (m)
A	2675.454	3748.621
B	2845.736	4972.814
C	3047.162	3796.734
D	2965.131	5087.163

A curve of radius 400 m is to be set out to join the straights AB and CD. Calculate (a) the coordinates of the intersection point I; (b) the coordinates of the tangent points; (c) the length and bearing of the long chord.

(a) To find the coordinates of the intersection point, by equation 5.31,

$$\Delta N_{ai} = (\Delta E_{ab} - n \Delta N_{ab})/(m - n)$$
$$\tan \phi_{ab} = m = \Delta E_{ab}/\Delta N_{ab}$$
$$= 170.282/1224.187 = 0.139\,098$$
$$\phi_{ab} = 7.9189°$$
$$\tan \phi_{cd} = n = \Delta E_{cd}/\Delta N_{cd}$$
$$= -82.031/1290.429 = -0.063\,569$$
$$\phi_{cd} = 356.3627°$$

The deflection is given by

$$\Delta = \phi_{dc} - \phi_{ab}$$
$$= 176.3627° - 7.9189° = 168.4438°$$

then

$$\Delta N_{ai} = [371.708 - (-0.063\,569 \times 48.113)]/(0.139\,098 + 0.063\,569)$$
$$= 1849.174 \text{ m}$$

$$\Delta E_{ai} = m \Delta N_{ai} = 0.139\,098 \times 1849.174$$
$$= 257.216 \text{ m}$$

Thus $E_i = 2932.670$ m and $N_i = 5597.795$ m

(b) The tangent lengths are given by

$$R \tan(\Delta/2) = 400 \tan 84.2219 = 3952.954 \text{ m}$$

To compute the coordinates of the tangent points.
For line IT_1,

$$\begin{aligned} S &= 3952.954 \text{ m} & \phi &= 187.9189° \\ \Delta E &= -544.604 \text{ m} & E_{T1} &= 2388.066 \text{ m} \\ \Delta N &= -3915.259 \text{ m} & N_{T1} &= 1682.536 \text{ m} \end{aligned}$$

For line IT_2

$$\begin{aligned} S &= 3952.954 \text{ m} & \phi &= 176.3627° \\ \Delta E &= 250.776 \text{ m} & E_{T2} &= 3183.446 \text{ m} \\ \Delta N &= -3944.991 \text{ m} & N_{T2} &= 1652.804 \text{ m} \end{aligned}$$

(c) To find the length and bearing of the long chord,

$$\phi_{T1/T2} = \tan^{-1}(795.380/-29.732) = 92.1408°$$

Check. $\quad \Delta = 2(\phi_{T1/T2} - \phi_{T1/I})$
$$= 2(092.1408 - 7.9189) = 168.4438° \quad \text{as above}$$

The length is

$$T_1 T_2 = 795.380 \text{ cosec } 92.1408 = 795.936 \text{ m}$$

5.8 Setting out of curves

5.8.1 By linear equipment only

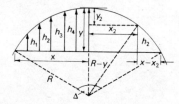

Fig. 5.12

(a) Offsets from the long chord (see Fig. 5.12).

If, in Fig. 5.12, the chord is subdivided into an even number of equal parts, the offsets h_1, h_2 etc. can be set out. Each side of the midpoint will be symmetrical. As
$$(R-y)^2 = R^2 - x^2$$
$$y = R - (R^2 - x^2)^{1/2} \tag{5.33}$$
$$y_2 = R - (R^2 - x_2^2)^{1/2}$$
$$h_2 = y - y_2$$

Note that if Δ is known,

$$y = R(1 - \cos \Delta/2) \tag{5.34}$$

Example 5.11 A kerb is part of a 100 m radius curve. If the chord joining the tangent points is 60.000 m long, calculate the offsets from the chord at 10 m intervals.

By equation 5.33,

$$h_3 = y = R - (R^2 - x^2)^{1/2}$$
$$= 100.000 - (100.000^2 - 30.000^2)^{1/2} = 4.606 \text{ m}$$
$$h_2 = h_3 - y_2$$

Therefore

$$y_2 = 100.000 - (100.000^2 - 10.000^2)^{1/2} = 0.501 \text{ m}$$
$$h_2 = 4.606 - 0.501 = 4.105 \text{ m}$$

Therefore

$$y_1 = 100.000 - (100.000^2 - 20.000^2)^{1/2} = 2.020 \text{ m}$$
$$h_1 = 4.606 - 2.020 = 2.586 \text{ m}$$

(b) Offsets from the straight (see Fig. 5.13).

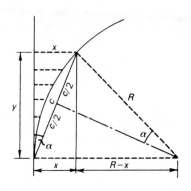

Fig. 5.13

As above,

$$(R - x)^2 = R^2 - y^2$$

Therefore $\quad x = R - (R^2 - y^2)^{1/2} \quad$ (5.35)

Alternatively,

$$\sin \alpha = c/2R = x/c$$

Therefore $\quad x = c^2/2R \quad$ (5.36)

If α is small, $c \simeq y$ and

$$x \simeq y^2/2R \quad (5.37)$$

(c) Offsets from the bisection of the chord (see Fig. 5.14).
By equation 5.35,

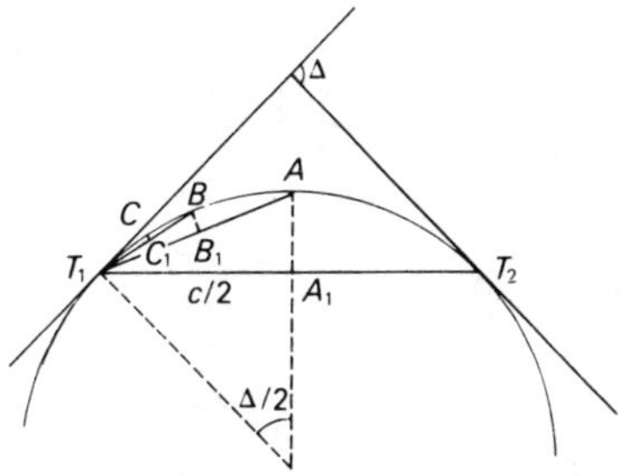

$$AA_1 = R - [R^2 - (c/2)^2]^{1/2} \simeq c^2/2R$$

If $T_1A \simeq T_1A_1 = c/2$, then

$$BB_1 = (c/2)^2/2R = c^2/8R = AA_1/4$$

Alternatively, by equation 5.34,

$$AA_1 = R(1 - \cos \Delta/2)$$
$$BB_1 = R(1 - \cos \Delta/4) \quad \text{etc.}$$

Fig. 5.14

(d) Offsets from the bisection of successive chords (see Fig. 5.15).
As above,

$$\sin \alpha = c/2R$$

Assuming equal chords,

$$T_1 t = Aa = Bb = Dd \quad \text{etc}$$
$$= R(1 - \cos \alpha)$$
$$= R - (R^2 - c^2/4)^{1/2}$$

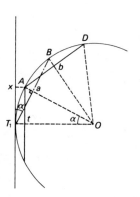

Fig. 5.15

200 CIRCULAR CURVES

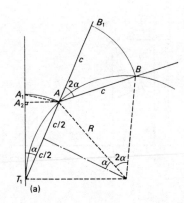

Fig. 5.16(a)

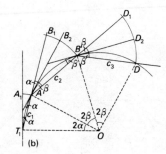

Fig. 5.16(b)

Lay off $T_1 x = c/2$

Lay off $xA = T_1 t$

$Aa = T_1 t$ along the line AO

$T_1 B$ on the line Ta produced

$Bb = T_1 t$ along line BO.

(e) Offsets from chords produced.

(i) Equal chords (Fig. 5.16(a)).

$$A_2 A = c \sin \alpha = c^2/2R \tag{5.38}$$

If α is small ($c \leq R/20$), then

$$A_2 A \simeq A_1 A$$
$$B_1 B \simeq 2 A_1 A = c^2/R$$

(ii) Unequal chords (through chainage) (Fig. 5.16(b)).

Offset from subchord is

$$O_1 = A_1 A = c_1^2/2R \tag{5.39}$$

Offset from the first full chord is

$$\begin{aligned} O_2 &= B_1 B = B_1 B_2 + B_2 B \\ &= (O_1 \times c_2/c_1) + c_2^2/2R \\ &= c_1 c_2 / 2R + c_2^2/2R \\ &= c_2(c_1 + c_2)/2R \end{aligned} \tag{5.40}$$

Offset from the second full chord is

$$\begin{aligned} O_3 &= D_1 D = D_1 D_2 + D_2 D \\ &= 2 D_2 D = 2 B_2 B = c_2^2/R \end{aligned}$$

Then $O_3 = c_3(c_2 + c_3)/2R = c_2^2/R$

Generally, the equation becomes

$$O_n = c_n(c_n + c_{n-1})/2R \tag{5.41}$$

5.8.2 By linear and angular equipment

(a) Tangential deflection angles (see Fig. 5.16(a)).

By equation 5.38,

$$\sin \alpha = c/2R$$

If α is small ($c \leq R/20$), then

$$\alpha_{\text{rad}} = c/2R$$

$$\alpha'' = 206\,265\,c/2R = kc \tag{5.42}$$

$$\alpha° = 57.2958\,c/2R = 28.6479\,c/R = Kc \tag{5.43}$$

For equal chords, $2\alpha = \Delta/n$ where n is the number of chords required

For sub-chords,

$$\alpha_1 = \frac{\alpha_s \times \text{sub-chord } (c_1)}{\text{standard chord } (c_s)} \qquad (5.44)$$

(b) Deflection angles from the chord produced (see Fig. 5.17).

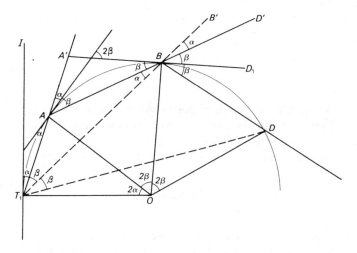

Fig. 5.17

Because of an obstruction the theodolite may have to be moved from the tangent point on to the established points on the curve.

(i) If the chords are equal then the deflection angles become

$\alpha, 2\alpha, 3\alpha, 4\alpha, \ldots n\alpha$

(ii) If the chords are not equal then the deflection angles become

$\alpha, (\alpha + \beta), (\alpha + 2\beta), (\alpha + 3\beta)$ etc.

These are the same values as the total deflection angle from the tangent point for each point on the curve, i.e.

Angle $A'AB = (\alpha + \beta) = IT_1 B$
Angle $B'BD = (\alpha + 2\beta) = IT_1 D$

To establish the tangent to B, a 'back angle' $T_1 BA' = (\alpha + \beta)$ may be set out and with the theodolite transitted, the line BD, is established (see Transition Curves, p. 236).

5.8.3 By angular equipment only (Fig. 5.18)

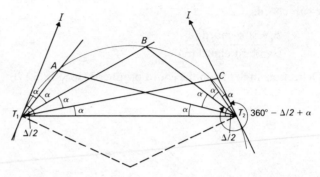

Fig. 5.18

By the use of two theodolites, points on the curve are cut in from the tangent points; for example, assuming equal chords A is the intersection of α from T_1 and 3α from T_2. As the deflection angle is not usually quoted as a deflection left it would be given as

$$360 - 3\alpha \quad \text{or} \quad 360 - \Delta/2 + \alpha \tag{5.45}$$

Example 5.12 A curve of radius 400 m is deflected $35°\,40'\,00''$ at the intersection point I of chainage 1234.260 m. Calculate (a) the tangent length; (b) the chainages of T_1 and T_2; (c) the deflection angles from the tangent point T_1 up to peg D of chainage 1160.000 m; (d) the 'back angle' at D; (e) the deflection angle at D to T_2.

(a) Tangent length is given by

$$R \tan \Delta/2 = 400.000 \tan 17.8333° = 128.683 \text{ m}$$

(b) Chainage of T_1 is given by

$$\text{chainage of } I - TL = 1234.260 - 128.683$$
$$= 1105.577 \text{ m}$$

Length of arc $T_1 T_2$ is

$$R \Delta_{\text{rad}} = 400.000 \times 35.6667/57.2938 = 249.000 \text{ m}$$
$$\text{chainage of } T_2 = 1354.577 \text{ m}$$

(c) Deflection angles (α). Assuming $c_s \leqslant R/20$, say 20 m,

$$\alpha° = 57.2958\, c/2R$$
$$= [(57.2958)/(2 \times 400)]\, c = 0.0716c$$
$$c_1 = 1120.000 - 1105.577 \qquad \alpha_1° = 0.0716 \times 14.423$$
$$= 14.423 \text{ m} \qquad\qquad\qquad = 1.0330°$$

$$c_s = 20.000 \text{ m} \qquad \alpha_s^\circ = 0.0716 \times 20.000$$
$$= 1.4324^\circ$$
$$c_n = 1354.577 - 1340.000 \qquad \alpha_n^\circ = 0.0716 \times 14.577$$
$$= 14.577 \text{ m} \qquad = 1.0440$$

Tabulating,

Point	Chainage	c	α°	$\Sigma\alpha^\circ$	Angle set out (δ)
T_1	1105.577				
a	1120.000	14.423	1.0330	1.0330	1° 01' 59"
b	1140.000	20.000	1.4324	2.4654	2° 27' 55"
c	1160.000	20.000	1.4324	3.8977	3° 53' 52"
$n-1$	1340.000	20.000	1.4324	16.7893	16° 47' 21"
T_2	1354.577	14.577	1.0440	17.8333	17° 50' 00"

(d) The back angle is then $\alpha = 3° 53' 52''$.

(e) The deflection angle to T_2 is

$$\delta_{\text{rad}} = \text{remaining arc}/2R = \frac{1354.577 - 1160.000}{800.000}$$
$$= 0.2432$$

Thus $\delta = 0.2432 \times 57.2958 = 13.9356°$

Check. Total deflection from $T_1 = 3.8977 + 13.9356$
$= 17.8333°$.

Exercises 5.3

1 A curve of radius 500 m is to be set out to join two straights BI and IC which deflect through 120° 00' 00". Point I is inaccessible and the following data was measured: BC = 525.200 m, angle IBC = 65° 30' 00", angle BCI = 54° 30' 00". Assuming the chainage of B is 23 + 23.65 m calculate sufficient data to set out the curve by deflection angles from the tangent point T_1 by standard chords of 20 m, based upon through chainage.
(TP Ans. 866.03 m; deflection angles, 1786", 4125", 3824")

2 The tangent length of a simple curve was 663.14 m and the deflection angle for a 100 m chord 2° 18' 00". Calculate the radius, the total deflection angle, the length of the curve and the final deflection angle.
(LU Ans. 1245.5 m; 56° 03' 50", 1218.7 m; 28° 01' 55")

3   A right hand circular curve of 500 m radius is to be set out to connect two straights AI and IB with bearings of 136° 51' and 161° 09' respectively. The forward chainage of I is 1239.80 m and pegs are to be placed along the curve at through chainages of 20 m. Tabulate the data required to set out the curve from the first tangent point with a theodolite reading to 20".

If it is subsequently found that (a) from the first tangent point it is impossible to sight any pegs on the curve beyond the third but from this point all the remaining points may be sighted and (b) owing to a local obstruction it is impossible to chain between pegs 5 and 6, explain the procedure you would adopt and calculate any further data you would require in setting out the curve.   (LU Ans. 0° 27' 00", 1° 35' 40")

4   A control station A is fixed for the setting out of a circular curve T_1T_2 with an intersection point I. Given the data below compute (a) the coordinates of T_2; (b) the polar coordinates from A to T_1 and T_2; (c) the setting out data for the deflection angles from T_1 for 10 m chords based upon through chainage.

	E (m)	N (m)	
T_1	1076.00	1092.00	chainage 943.20 m
A	1095.30	1109.40	

Bearing T_1I 067° 15' 00"
Bearing IT_2 117° 15' 00"
Radius   100.00 m
(RICS/M Ans. (a) 1160.46 m E, 1088.68 m N; (b) 25.99 m, 227° 57' 49"; 68.39 m, 107° 38' 03"; (c) 1° 56' 53", 2° 51' 53", 0° 08' 04")

5   AB and CD are two straights bearing 065° 30' and 249° 45' respectively which are to be connected by two separate arcs BE radius 660 m and FD 990 m radius, EF being the straight between the arcs. Given the following coordinate values,

	E (m)	N (m)	
B	8435.80	10 603.90	chainage 3624.20 m
D	11 108.80	9663.40	

Calculate (a) the coordinates of E and F; (b) the chainages of E, F and D.
(RICS/M Ans. (a) 9083.24 m E, 10 547.31 m N; 10 205.53 m E, 9776.24 m N; (b) 4303.73 m, 5665.36 m, 6611.22 m)

5.9 Computation of coordinates around the curve

In large scale operations, e.g. motorways, curves are frequently set out from control stations by either (a) polar coordinates using one theodolite and EDM or (b) two theodolites (see Fig. 5.10). In either case, the coordinates of the points on the curve must be computed. Two methods are described below.

5.9.1 By polar coordinates from the centre of the curve

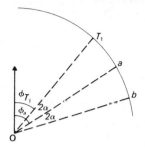

Fig. 5.19

With reference to Fig. 5.19,

$$\phi_a = \phi_{T1} + 2\alpha$$

where α is the normal deflection angle from the tangent,

$$E_a = E_0 + R \sin \phi_a$$
$$N_a = N_0 + R \cos \phi_a$$

Note that

$$2\alpha_{\text{rad}} = \text{arc}_{T1/a}/R$$
$$2\alpha = 57.2958 l/R$$

where $l = \text{chainage}_a - \text{chainage}_{T1} = C_a - C_{T1}$.

Also, if ϕ is changed into radians, then by setting the calculator into radian mode, the coordinates are derived by total commitment to radians. The coordinates of the centre (O) must be initially obtained.

5.9.2 By rectangular coordinates from the tangent point T_1

With reference to Fig. 5.20, as before

$$l = C_p - C_{T1}$$

and $2\alpha = l/R$

$$x = R \sin 2\alpha \tag{5.46}$$
$$y = R(1 - \cos 2\alpha) \tag{5.47}$$

Note that y is negative for a right hand curve.

To find the coordinates of P, it is necessary to transform the cartesian coordinates (x, y) into the rectangular coordinates (E_p, N_p). By equations 2.34 and 2.35,

$$E_p = E_{T1} + mx - ny$$
$$N_p = N_{T1} + my + nx$$

where $m = \cos \theta$, $n = \sin \theta$ and $\theta = \text{old } \phi - \text{new } \phi$
$$= 90° - \phi_{T1/l}.$$

206 CIRCULAR CURVES

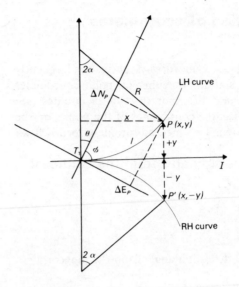

Fig. 5.20

Example 5.13 Given the centre of the curve as 1000.000 m E, 1000.000 m N and $R = 400$ m; $\phi_{O/T1} = 30°\,00'\,00''$; chainage of $T_1 = 123.461$ m; $c_s = 20$ m.

Method 1 (coordinates from centre of curve).
For line OT_1

$$S = R = 400.000 \text{ m} \quad \phi = 30.0000°$$
$$\Delta E = 200.000 \text{ m} \quad E_{T1} = 1200.000 \text{ m}$$
$$\Delta N = 346.410 \text{ m} \quad N_{T1} = 1346.410 \text{ m}$$

To obtain the coordinates of P (chainage 140.000 m)

$$2\alpha_{\text{rad}} = l/R = (C_P - C_{T1})/R$$
$$= (140.000 - 123.461)/400 = 0.041\,35$$
$$2\alpha = (57.2958 \times 16.539)/400 = 2.3690°$$

$$\phi_{O/T1} = 30.0000$$
$$\phi_{O/a} = 30.0000 + 2.3690 = 32.3690$$

$$\Delta E = 400 \sin 32.3690 = 214.48 \quad E_p = 1214.48$$
$$\Delta N = 400 \cos 32.3690 = 337.847 \quad N_p = 1337.847$$

Method 2 (coordinates from the tangent point T_1)

Using the values above and equations 5.46 and 5.47,

$$x = R \sin 2\alpha = 400 \sin 2.3690 = 16.5340$$
$$y = R(1 - \cos 2\alpha) = -400(1 - \cos 2.3690)$$
$$= -0.3419$$
$$\theta = 90.0000° - \phi_{T1/I}$$
$$= 90.0000° - 120.0000° = -30.0000°$$

5.9 COMPUTATION OF COORDINATES AROUND THE CURVE

Then $\quad E_p = E_{T1} + mx - ny$
where $\quad m = \cos\theta = 0.86603$ and $n = \sin\theta = -0.500000$.

$$E_p = 1200.000 + (0.86603 \times 16.5340)$$
$$\quad - (-0.50000 \times -0.3419)$$
$$= 1214.148 \text{ m as above}$$
$$N_p = N_{T1} + my + nx$$
$$= 1346.410 + (0.86603 \times -0.3419)$$
$$\quad + (-0.50000 \times 16.5340)$$
$$= 1337.847 \text{ m as above}$$

Example 5.14 Given $\phi_{T1/I} = 25°00'00''$; $\Delta = 30°00'00''$; $R = 200$ m; coordinates of $I = 1000.000$ m E, 1000.000 m N; chainage 1236.420 m; compute the setting out data including the coordinates of the first five pegs for a right hand curve.

Size of chord is

$$c_s \leqslant R/20 = 10 \text{ m}$$

Tangent length is

$$R \tan \Delta/2 = 200.000 \tan 15.0000° = 53.590 \text{ m}$$

For line IT_1,

$S = 53.590$ m $\qquad \phi = (180 + 25.0000)° = 205.0000°$
$\Delta E = -22.648$ m $\quad E_{T1} = 977.352$ m
$\Delta N = -48.569$ m $\quad N_{T1} = 951.431$ m

For line IT_2,

$S = 53.590$ m $\qquad \phi = (25.0000 + 30.0000)°$
$\qquad\qquad\qquad\qquad = 55.0000°$
$\Delta E = 43.898$ m $\quad E_{T2} = 1043.898$ m
$\Delta N = 30.738$ m $\quad N_{T2} = 1030.738$ m

Chainages

$I = 1236.420$ m
$TL = \underline{53.590 \text{ m}}$
$T_1 = 1182.830$ m
Arc $R\Delta_{\text{rad}} = \underline{104.720 \text{ m}}$
$T_2 = 1287.550$ m

To compute the coordinates of O, for line T_1O,

$S = R = 200.000$ m $\qquad \phi = (25.0000 + 90.0000)°$
$\qquad\qquad\qquad\qquad\qquad = 115.0000°$
$\Delta E = 181.262$ m $\qquad E_O = 1158.614$ m
$\Delta N = -84.524$ m $\qquad N_O = 866.907$ m

Bearing $OT_1 = 295.0000°$

Tabulating,

Point	Chainage	l	2α	ϕ	E	N	Setting out angle (δ)
T_1	1182.830			295.0000°	977.352	951.431	
a	1190.000	7.170	2.0540°	297.0540°	980.498	957.873	1° 01' 37"
b	1200.000	17.170	4.9188°	299.9188°	985.267	966.662	2° 27' 34"
c	1210.000	27.170	7.7836°	302.7836°	990.469	975.201	3° 53' 30"
d	1220.000	37.170	10.6484°	305.6484°	996.092	983.469	5° 19' 27"
T_2	1287.550	104.720	30.0000°	325.0000°	1043.898	1030.738	15° 00' 00"

5.10 Highway alignment design

The desired alignment is usually drawn on a plan with the aid of railway curves and the subsequent computation is dependent upon varying degrees of freedom involving three types of elements.

1) Fixed—precisely located in position.
2) Floating—passing through a fixed point to permit rotation.
3) Free—with no fixed parameters.

The design must proceed in steps, two elements at a time, i.e. the second is fixed relative to the first, etc.

Fixed elements are

a) a straight line ($R = \infty$), in which two points are fixed or in which one is a fixed point + a bearing;
b) a fixed curve passing through (i) 3 fixed points *or* (ii) 2 fixed points + radius *or* (iii) 1 fixed point + radius + bearing of the straight *or* (iv) centre coordinates + radius or D.

Floating elements are

a) a straight passing through a fixed point or of a given bearing;
b) a curve passing through a fixed point + a given radius.

Free elements are

a) a straight with no fixed points;
b) a curve of given radius or D.

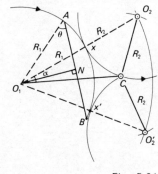

Fig. 5.21

Example 5.15 (Floating element) A fixed curve passing through A and B is required to pass through C (Fig. 5.21). The radii of the required curves are $R_1 = 200$ m(RH) and $R_2 = 100$ m(LH).

Coordinates of the fixed points are as follows.

	E (m)	N (m)
A	1000.000	1000.000
B	1208.000	775.000
C	1220.000	850.000

The centre O_2 must lie on the arc with centre C and radius R_2 and also the arc with centre O_1 and radius $(R_1 + R_2)$. Two solutions are possible, O_2 and O_2', with common tangent points X and X'. Then

$$\phi_{ab} = \tan^{-1}(208.000/-225.000) = 137.2483°$$
$$S_{ab} = (208.000^2 + 225.000^2)^{1/2} = 306.413 \text{ m}$$
$$S_{ab}/2 = 153.207 \text{ m}$$

In $\triangle O_1 AN$,

$$\theta = \cos^{-1}(AN/R_1) = 40.0008°$$
$$\phi_{a/01} = \phi_{ab} + \theta$$
$$= 137.2483° + 40.0008° = 177.2491°$$

To obtain the coordinates of O_1, for line AO_1

$$S = R_1 = 200.000 \text{ m} \quad \phi = 177.2491°$$
$$\Delta E = 9.599 \text{ m} \quad E_{O1} = 1009.599 \text{ m}$$
$$\Delta N = -199.769 \text{ m} \quad N_{O1} = 800.231 \text{ m}$$
$$\phi_{O1/c} = \tan^{-1}(210.401/49.769) = 76.6917°$$
$$S_{O1/c} = (210.401^2 + 49.769^2)^{1/2} = 216.207 \text{ m}$$

In triangle $O_1 O_2 C$,

$$\cos O_1 = [(R_1 + R_2)^2 + O_1C^2 - O_2C^2]/[2 \times (R_1 + R_2) \times O_1C]$$
$$= (300^2 + 216.207^2 - 100^2)/(2 \times 300 \times 216.207)$$
$$O_1 = 12.3020°$$
$$\phi_{O1/O2} = \phi_{O1/c} - O_1$$
$$= 76.6917° - 12.3020° = 64.3897°$$

To obtain the coordinates of X, for line $O_1 X$

$$S = 200.000 \text{ m} \quad \phi = 64.3897°$$
$$\Delta E = 180.351 \text{ m} \quad E_x = 1189.950 \text{ m}$$
$$\Delta N = 86.540 \text{ m} \quad N_x = 886.681 \text{ m}$$

To obtain the coordinates of O_2, for line $O_1 O_2$

$$S = 300.000 \text{ m} \quad \phi = 64.3897°$$
$$\Delta E = 270.526 \text{ m} \quad E_{O2} = 1280.125 \text{ m}$$
$$\Delta N = 129.675 \text{ m} \quad N_{O2} = 929.906 \text{ m}$$

Note that the alternative solution would be obtained by adding the angle $O_1 = 12.3020°$ to the bearing $O_{O1/c}$ and then continuing as above.

Exercise 5.4

1 Given the data below and using Fig. 5.22, calculate the coordinates of O_1, O_2, O_3, X_1, and X_2.

Station	E (m)	N (m)	
1	1065.000	852.500	Chainage 100.000 m
2	1200.000	1085.000	
3	920.000	1022.000	
4	900.000	877.000	

$R_1 = 190\,\text{m}$; $R_2 = 74\,\text{m}$; $R_3 = -103\,\text{m}$ (i.e. left hand).

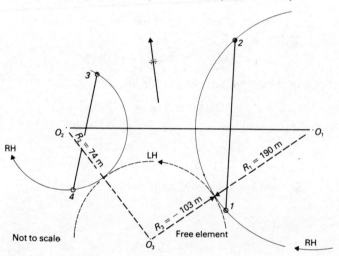

Fig. 5.22

(Ans. 1248.619 m E, 901.326 m N; 899.160 m E, 950.995 m N; 976.804 m E, 791.933 m N; 1072.358 m E, 830.389 m N, 931.621 m E, 884.495 m N)

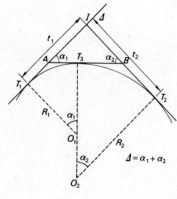

Fig. 5.23

5.11 Compound curves

Compound curves consist of two or more consecutive circular arcs of different radii, having their centres on the same side of the curve. There are seven components in a compound curve made up of two arcs; two radii, R_1 and R_2; two tangent lengths, t_1 and t_2; an angle subtended by each arc, α_1 and α_2; and a deflection angle at the intersection point I, $\Delta = \alpha_1 + \alpha_2$. At least four values must be known.

These seven elements may be related in the following equations.

$$t_1 \sin \Delta = (R_2 - R_1) \text{ versine } \alpha_2 + R_1 \text{ versine } \Delta \quad (5.48)$$
$$t_2 \sin \Delta = (R_1 - R_2) \text{ versine } \alpha_1 + R_2 \text{ versine } \Delta \quad (5.49)$$

The equations may be proved by the following construction (see Fig. 5.24(a)).

1) Produce arc $T_1 T_3$ of radius R_1 to B.
2) Draw $O_1 B$ parallel to $O_2 T_2$, BC parallel to IT_2, BD parallel to $T_3 O_2$, $T_1 A$ perpendicular to $O_1 B$, $T_1 E$ perpendicular to $T_2 I$ (produced).

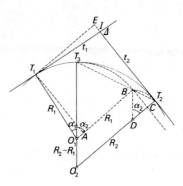

Fig. 5.24(a)

Note that $T_3 B T_2$ and $T_3 O_1 O_2$ are straight lines, and that

$$BD = O_1 O_2 = DT_2 = R_2 - R_1 \quad (R_2 > R_1)$$

Then $\quad T_1 E = AB + CT_2$
$$= (O_1 B - O_1 A) + (DT_2 - DC)$$

i.e. $t_1 \sin \Delta = R_1 - R_1 \cos(\alpha_1 + \alpha_2) + [(R_2 - R_1)$
$$- (R_2 - R_1) \cos \alpha_2]$$
$$= R_1(1 - \cos \Delta) + (R_2 - R_1)(1 - \cos \alpha_2)$$
$$t_1 \sin \Delta = (R_2 - R_1) \text{ versine } \alpha_2 + R_1 \text{ versine } \Delta$$
(equation 5.48)

By similar construction (Fig. 5.24(b)),

$$FT_2 = GH - HJ$$
$$= (O_2 H - O_2 G) - (HK - JK)$$
$$t_2 \sin \Delta = (R_1 - R_2) \text{ versine } \alpha_1 + R_2 \text{ versine } \Delta$$
(equation 5.49)

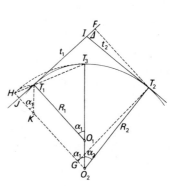

Fig. 5.24(b)

5.11.1 Alternative solution by using the 'equivalent circle'

In Fig. 5.25 $T_1 I$, AB, and IT_2 are tangents common to the arcs $T_1 T_3$ (radius R_1), $T_3 T_2$ (radius R_2), and to the assumed 'equivalent' arc KMN (radius R). Then

$$KT_1 = T_3 M = (R - R_1) \tan \alpha_1/2 \quad (5.50)$$
$$NT_2 = T_3 M = (R_2 - R) \tan \alpha_2/2 \quad (5.51)$$
$$(R - R_1) \tan \alpha_1/2 = (R_2 - R) \tan \alpha_2/2$$
$$R(\tan \alpha_1/2 + \tan \alpha_2/2) = R_2 \tan \alpha_2/2 + R_1 \tan \alpha_1/2$$

The 'equivalent radius' is given by
$$R = (R_1 \tan \alpha_1/2 + R_2 \tan \alpha_2/2)/(\tan \alpha_1/2 + \tan \alpha_2/2) \quad (5.52)$$

and $T_1 I = t_1 = KI - T_3 M$
$= R \tan \Delta/2 - (R - R_1) \tan \alpha_1/2$
$= R \tan \Delta/2 - (R_2 - R) \tan \alpha_2/2$
$T_2 I = t_2 = NI + T_3 M$
$= R \tan \Delta/2 + (R - R_1) \tan \alpha_1/2 = R \tan \Delta/2$
$\quad + (R_2 - R) \tan \alpha_2/2$

Then $(t_1 + t_2)/2 = R \tan \Delta/2$

$$R = (t_1 + t_2)/(2 \tan \Delta/2) \tag{5.53}$$

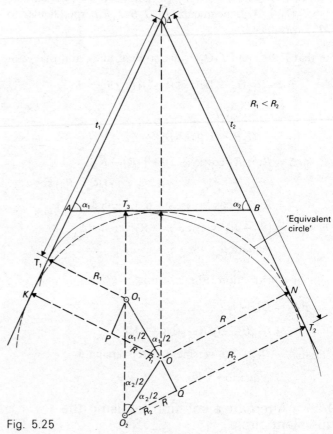

Fig. 5.25

Also $(t_2 - t_1)/2 = (R - R_1) \tan \alpha_1/2$
$$= (R_2 - R) \tan \alpha_2/2 \tag{5.54}$$

Substituting for R (equation 5.53) gives

$$(t_2 - t_1)/2 = \left[(t_1 + t_2)/(2 \tan \Delta/2) - R_1\right] \tan \alpha_1/2 \tag{5.55}$$
$$= \left[R_2 - (t_1 + t_2)/(2 \tan \Delta/2)\right] \tan \alpha_2/2 \tag{5.56}$$
$$\tan \alpha_1/2 = \tfrac{1}{2}(t_2 - t_1)/(R - R_1) \tag{5.57}$$
$$\tan \alpha_2/2 = \tfrac{1}{2}(t_2 - t_1)/(R_2 - R) \tag{5.58}$$
$$R_2 = \tfrac{1}{2}(t_2 - t_1)/\tan \alpha_2/2 + R \tag{5.59}$$

It is necessary to choose the appropriate equations for the data given.

5.11 COMPOUND CURVES

Example 5.16 Given $R_1 = 20$ m, $R_2 = 40$ m, $T_1 I = 20.50$ m, $\Delta = 80°\ 30'$, calculate the tangent length $T_2 I$.

By equation 5.48,

$$\text{versine } \alpha_2 = (T_1 I \sin \Delta - R_1 \text{ versine } \Delta)/(R_2 - R_1)$$
$$= [20.50 \sin 80.50 - 20.00(1 - \cos 80.50)]/20.00$$
$$= 0.176$$

then $\alpha_2 = 34.51°$

$\alpha_1 = \Delta - \alpha_2 = 80.50 - 34.51 = 45.99°$

By equation 5.49,

$$T_2 I = (R_2 \text{ vers } \Delta - (R_2 - R_1) \text{ vers } \alpha_1)/\sin \Delta$$
$$= [40.00(1 - \cos 80.50°) - 20.00(1 - \cos 45.99°)]/\sin 80.50°$$
$$= 27.67 \text{ m}$$

5.11.2 Alternative solution from first principles (Fig. 5.26)

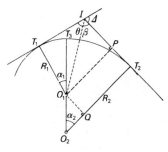

Fig. 5.26

Construction. Join $O_1 I$. Draw $O_1 P$ parallel to $O_2 T_2$ and $O_1 Q$ parallel to IT_2. Then in triangle $T_1 I O_1$

$$\theta = \tan^{-1}(R_1/T_1 I) = \tan^{-1}(20.00/20.50) = 44.29°$$
$$\beta = 180° - (\theta + \Delta) = 180 - (44.29 + 80.50) = 55.21°$$

In triangle IPO_1,

$$IP = O_1 I \cos \beta = R_1 \cos \beta / \sin \theta$$
$$= 20.00 \cos 55.21 / \sin 44.29 = 16.34 \text{ m}$$
$$O_1 P = O_1 I \sin \beta = R_1 \sin \beta / \sin \theta = 23.52 \text{ m}$$
$$O_2 Q = O_2 T_2 - QT_2 = R_2 - O_1 P = 40.00 - 23.52$$
$$= 16.48 \text{ m}$$

In triangle $O_1 O_2 Q$,

$$\alpha_2 = \cos^{-1}(O_2 Q / O_1 O_2) = \cos^{-1}(16.48/20.00) = 34.51°$$
$$= 34°31'$$
$$\alpha_1 = \Delta - \alpha_2 = 80.50 - 34.51 = 45.99° = 45°59'$$
$$O_1 Q = O_1 O_2 \sin \alpha_1 = 20.00 \sin 34.51 = 11.33 \text{ m}$$
$$IT_2 = IP + PT_2 = IP + O_1 Q$$
$$= 16.34 + 11.33 = 27.67 \text{ m}$$

Example 5.17 AB and DC are the centre lines of two straight portions of a railway which are to be connected by means of a compound curve BEC; BE is one circular curve and EC the other. The radius of the circular curve BE is 200 m. Given the coordinates in metres, B 100.00 m E, 200.00 m N; C 268.00 m E, 296.50 m N. Bearing $AB = 025°\ 30'$, $DC = 283°\ 30'$. Calculate

(a) the coordinates of E; (b) the radius of the circular curve EC.
(RICS/M)

$$\phi_{bc} = \tan^{-1}(268.00 - 100.00)/(296.50 - 200.00)$$
$$= 60.13°$$
$$S_{bc} = 168.00/\sin 60.13° = 193.74 \text{ m}$$
$$\Delta = \phi_{cd} - \phi_{ab} = 103.50° - 25.50° = 78.00°$$

In triangle BIC,

$$B = 60.13° - 25.50° = 34.63°$$
$$C = 78.00° - 34.63° = 43.37°$$
$$BI = t_1 = BC \sin C / \sin \Delta$$
$$= 193.74 \sin 43.37°/\sin 78.00° = 136.01 \text{ m}$$
$$CI = t_2 = BC \sin B / \sin \Delta$$
$$= 193.74 \sin 34.63°/\sin 78.00° = 112.56 \text{ m}$$

Then $(t_1 + t_2)/2 = 124.285$ m

$$(t_2 - t_1)/2 = -11.725 \text{ m}$$

The 'equivalent radius' is given by

$$R = (t_1 + t_2)/(2 \tan \Delta/2) = 153.479 \text{ m (equation 5.55)}$$
$$\alpha_1 = 2 \tan^{-1}[(t_2 - t_1)/2(R - R_1)] = 28.292°$$
$$\alpha_2 = \Delta - \alpha_1 = 78.000 - 28.292 = 49.708°$$

Then $R_2 = (t_2 - t_1)/(2 \tan \alpha_2/2) + R$
$$= -11.725/\tan 24.854° + 153.479 = 128.16 \text{ m}$$

To find the coordinates of E, for line BE,

$$S = 2 \times 200.00 \sin \alpha_1/2 = 97.78 \text{ m}$$
$$\phi = 025.50 + 14.15 = 39.65°$$
$$\Delta E = 62.39 \text{ m} \qquad E_e = 162.39 \text{ m}$$
$$\Delta N = 75.29 \text{ m} \qquad N_e = 275.29 \text{ m}$$

Check. $t_1 \sin \Delta = 136.01 \sin 78.00° = 133.04$ m

$$R_1 \text{ vers } \Delta = 200.00(1 - \cos 78.00°) = 158.42 \text{ m}$$
$$(R_2 - R_1) \text{ vers } \alpha_2 = (128.15 - 200.00)(1 - \cos 49.70°)$$
$$= -25.38 \text{ m}$$
$$R_1 \text{ vers } \Delta + (R_2 - R_1) \text{ vers } \alpha_2 = 158.42 - 25.38$$
$$= 133.04 \text{ m} = t \sin \Delta$$

Example 5.18 Four points A, B, C and D are traverse control points having local coordinates as follows.

	E (m)	N (m)
A	146.58	539.74
B	290.52	1026.76
C	961.61	1054.88
D	1340.65	744.89

5.11 COMPOUND CURVES

A compound curve is to be set out between the tangent points A and D, the lines AB and CD being tangential to the curves. The initial radius (R_1) is 500 m. Assuming the chainage of the intersection point (I) is 2038.58 m, calculate: (1) the radius of the second arc (R_2); (2) the chainages of A and D; (3) the data required to set out the first five pegs beyond A on the first arc; (4) the back angle to establish the line of the common tangent at T_3. (TP)

1) $\tan \phi_{ab} = (290.52 - 146.58)/(1026.76 - 539.74)$
$= 0.295\,553 = m$
$\tan \phi_{dc} = (961.61 - 1340.65)/(1054.88 - 744.89)$
$= -1.222\,749 = n$
$\phi_{ab} = 16.465°$
$\phi_{dc} = 309.277°$
$\Delta = \phi_{cd} - \phi_{ab} = 129.277 - 16.465 = 112.812°$
$\Delta/2 = 56.406°$

To find the coordinates of the intersection point (I); by equation 5.31,

$\Delta N_{ai} = (\Delta E_{ad} - n\Delta N_{ad})/(m - n)$
$= [(1194.07 - (-1.222\,749 \times 205.15)]/(0.295\,553 + 1.222\,749)$
$= 951.67\,\text{m}$
$\Delta E_{ai} = \Delta N_{ai} \times m = 951.67 \times 0.295\,553 = 281.27\,\text{m}$

then $E_i = 427.85$ and $N_i = 1491.41\,\text{m}$

$t_1 = AI = (\Delta E_{ai}^2 + \Delta N_{ai}^2)^{1/2} = 992.37\,\text{m}$
$t_2 = DI = (\Delta E_{di}^2 + \Delta N_{di}^2)^{1/2} = 1179.19\,\text{m}$
$(t_1 + t_2)/2 = 1085.78\,\text{m}$
$(t_2 - t_1)/2 = 93.41\,\text{m}$

By equation 5.53,
$R = (t_1 + t_2)/(2 \tan \Delta/2) = 1085.78/\tan 56.406$
$= 721.23\,\text{m}$

By equation 5.57,
$\alpha_1 = 2 \tan^{-1}(t_2 - t_1)/[2(R - R_1)] = 45.782°$
$\alpha_2 = \Delta - \alpha_1 = 112.812 - 45.782 = 67.030°$

By equation 5.59,
$R_2 = (t_2 - t_1)/(2 \tan \alpha_2/2) + R$
$= 93.41/\tan 33.515 + 721.23 = 862.28\,\text{m}$

Length of arc
$AT_3 = R_1 \alpha_{1\text{rad}} = 500.00 \times 45.782/57.2958$
$= 399.52\,\text{m}$

Length of arc

$$T_3 D = R_2 \alpha_{2_{rad}} = 862.28 \times 67.030/57.2958$$
$$= 1008.78 \text{ m}$$

To find the chainages,

$$I = 2038.58 \text{ m}$$
$$t_1 = \underline{-992.37 \text{ m}}$$
$$A(T_1) = 1046.21 \text{ m}$$
$$\text{Arc } AT_3 = \underline{399.52 \text{ m}}$$
$$T_3 = 1445.73 \text{ m}$$
$$\text{Arc } T_3 D = \underline{1008.78 \text{ m}}$$
$$D(T_2) = 2454.51 \text{ m}$$

To set out the curve,

$$R = 500.00 \text{ m then } c_s \leqslant R/20 = \text{(say) } 25 \text{ m}$$

Point	Chainage	Chord	$\alpha°$	$\Sigma\alpha°$	Setting out angle (δ)
A	1046.21				
a	1050.00	3.79	0.2172°	0.2172°	0° 13′ 01″
b	1075.00	25.00	1.4324°	1.6496°	1° 38′ 58″
c	1100.00	25.00	1.4324°	3.0820°	3° 04′ 55″
d	1125.00	25.00	1.4324°	4.5144°	4° 30′ 52″
e	1150.00	25.00	1.4324°	5.9468°	5° 56′ 48″
T_3	1445.73	20.73	1.1877	22.8909	22° 53′ 27″

Check. $\alpha_1/2 = 45.782/2° = 22.891° = 22°53'27''$
The back angle $= \alpha_1/2 \qquad\qquad = 22° 53' 27''$

Exercises 5.5

1 A main haulage road AD, bearing due north, and a branch road DB, bearing 087° 00′, are to be connected by a compound curve formed by two circular curves of different radii in immediate succession. The first curve of 200 m radius starts from a tangent point A, 160.0 m due south of D, and is succeeded by a curve of 100 m radius which terminates at a tangent point on the branch road DB. Draw a plan of the roadways and the connecting curves to a scale of 1/500 and show clearly all construction lines. Thereafter, calculate the distance along the branch road from D to the tangent point of the second curve and the distance from A along the line of the first curve to the tangent point common to both curves.

(MQB/S Ans. 120.2 m, 145.4 m)

2 Undernoted are the coordinates in metres of points on the respective centre lines of two railway tracks ABC and DE.

	E	N
A	1000.000	1000.000
B	1160.118	1016.100
C	1252.298	1075.374
D	1003.319	966.996
E	1223.489	989.134

The lines AB and DE are straight and B and C are tangent points joined by a circular curve. It is proposed to connect the two tracks at C by a circular curve starting at X on the line DE, C being a tangent point common to both curves. Calculate the radius of each curve, the distance DX and the coordinates of X. (MQB/S Ans. 201.169 m, 120.701 m, 95.824 m, 1098.662 m E, 976.583 m N)

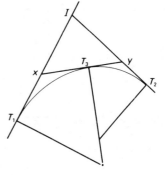

Fig. 5.27

3 A pit bottom layout is to consist of a compound curve $T_1T_3T_2$, (Fig. 5.27). Given the following data:

	E (m)	N (m)
T_1	550.36	426.87
I	583.26	481.77
T_2	631.06	449.17

Bearing $XT_3Y = 084°\,11'\,00''$ (common tangent)

calculate (a) the radii of the compound curve; (b) the length of each arc; (c) the distance XT_3; (d) the required data to show whether the centre line can be maintained by a single chord in each case assuming 3.00 m roadways.
(RICS/M Ans. (a) 63.55 m, 49.35 m (b) 59.06 m, 34.55 m
(c) 31.86 m, (d) second curve can be set out)

5.12 Reverse curves

There are four cases to consider.

1) Tangents parallel, radii equal or radii unequal.
2) Tangents not parallel, radii equal or radii unequal.

If the tangents are parallel (cross-overs, Fig. 5.28) then T_1T_2, O_1O_2 and I_1I_2 are all straight lines intersecting at E, the common tangent point. The construction is as follows.

Bisect T_1E and T_2E; draw perpendiculars PO_1 and QO_2. Then

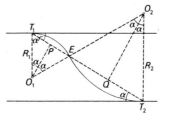

Fig. 5.28

$$T_1E = 2T_1P = 2R_1 \sin \alpha$$
$$T_2E = 2T_2Q = 2R_2 \sin \alpha = T_1T_2 - T_1E$$
$$= T_1T_2 - 2R_1 \sin \alpha$$

General equation

$$R_2 = (T_1T_2/2 \sin \alpha) - R_1 \qquad (5.60)$$

If $R_1 = R_2 = R = T_1T_2/(4 \sin \alpha)$ \qquad (5.61)

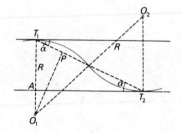

Example 5.19 Two parallel railway lines are to be connected by a reverse curve, each section having the same radius. If the centre lines are 30.00 m apart and the distance between the tangent points is 120.00 m, what will be the radius of the crossover?

$$T_1A = 30.00 \text{ m}$$
$$T_1T_2 = 120.00 \text{ m}$$

Fig. 5.29 In Fig. 5.29,

In triangle T_1T_2A

$$\sin \alpha = T_1A/T_1T_2 = 30.00/120.00$$

By equation 5.62,

$$R = T_1T_2/4 \sin \alpha = 120.00/4 \times 30/120 = 120.00 \text{ m}$$

If the tangents are not parallel and radii are equal, then the construction is as follows (Fig. 5.30).

Draw O_1S parallel to T_1T_2, and PO_1 perpendicular to T_1T_2, O_2Q perpendicular to T_1T_2. In triangle T_1PO_1,

$$T_1P = R \sin \alpha_1$$
$$PO_1 = R \cos \alpha_1$$

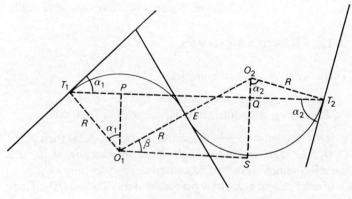

Fig. 5.30

In triangle O_2T_2Q,

$$T_2Q = R \sin \alpha_2 \qquad O_2Q = R \cos \alpha_2$$
$$O_2S = O_2Q + QS = R(\cos \alpha_1 + \cos \alpha_2)$$
$$\beta = \sin^{-1} O_2S/O_1O_2 = \sin^{-1} R(\cos \alpha_1 + \cos \alpha_2)/2R$$
$$O_1S = 2R \cos \beta = T_1T_2 - (T_1P + QT_2)$$
$$= T_1T_2 - R(\sin \alpha_1 + \sin \alpha_2)$$

Then $\quad R = T_1T_2/(2 \cos \beta + \sin \alpha_1 + \sin \alpha_2) \qquad (5.62)$

Example 5.20 Two underground roadways AB and CD are to be connected by a reverse curve of common radii with tangent points at B and C. If the bearings of the roadways are AB 096° 45′ and CD 105° 30′ and the coordinates of B 1125.66 m E, 1491.28 m N, and C 2401.37 m E, 650.84 m N, calculate the radius of the curve (Fig. 5.31).

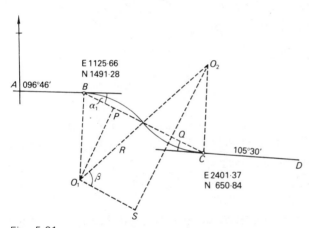

Fig. 5.31

The bearings of the tangents are different and thus the line BC does not intersect with O_1O_2 at the common tangent point although at this scale, the plotting might suggest this.

Bearing $BC = \tan^{-1}(2401.37 - 1125.66)/(650.84 - 1491.28)$
$\qquad = 123.377°$
Length $BC = (1275.71^2 + 840.44^2)^{1/2} = 1527.67$ m
Bearing $AB = 096.750 \quad \alpha_1 = 123.377 - 96.750 = 26.627°$
Bearing $CD = 105.500 \quad \alpha_2 = 123.377 - 105.500 = 17.877°$
$\qquad \beta = \sin^{-1}(\cos 26.627° + \cos 17.877°)/2 = 67.343°$
$\quad O_1S = 2R \cos 67.343°$
$\quad\; BP = R \sin \alpha_1 = R \sin 26.627°$
$\quad\; QC = R \sin \alpha_2 = R \sin 17.877°$
$\quad\; BC = BP + O_1S + QC$

$$1527.67 = R \sin 26.627° + 2R \cos 67.343°$$
$$+ R \sin 17.877°$$
$$R = 1527.67/(\sin 26.627° + 2 \cos 67.343° + \sin 17.877°)$$
(equation 5.62)
$$= 1001.37 \text{ m}$$

If tangents are not parallel and radii are not equal (Fig. 5.32), then the construction is as follows.

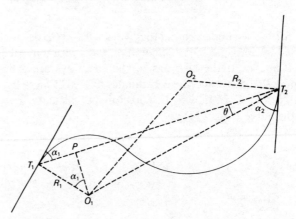

Fig. 5.32

Join O_1T_2. Draw O_1P perpendicular to T_1T_2. Then
$$T_1P = R_1 \sin \alpha_1$$
$$O_1P = R_1 \cos \alpha_1$$
$$PT_2 = T_1T_2 - T_1P = T_1T_2 - R \sin \alpha_1$$
$$\theta = \tan^{-1} O_1P/PT_2$$
$$= \tan^{-1}[(R_1 \cos \alpha_1)/(T_1T_2 - R \sin \alpha_1)] \quad (5.63)$$
$$O_1T_2 = O_1P/\sin \theta = R_1 \cos \alpha_1/\sin \theta$$

In triangle $O_1O_2T_2$,
$$O_1O_2{}^2 = O_2T_2{}^2 + O_1T_2{}^2 - 2\, O_2T_2 \cdot O_1T_2 \cos O_1T_2O_2$$
$$(R_1 + R_2)^2 = R_2{}^2 + [R_1{}^2 \cos^2 \alpha_1/\sin^2 \theta]$$
$$- [2R_2R_1 \cos \alpha_1 \cos(90° - (\alpha_2 - \theta)/\sin \theta]$$
$$R_1{}^2 + 2R_1R_2 + R_2{}^2 = R_2{}^2 + [\{R_1{}^2 \cos^2 \alpha_1$$
$$- 2R_1R_2 \cos \alpha_1 \sin(\alpha_2 - \theta)\sin \theta\}/\sin^2 \theta]$$

Then
$$R_2 = [R_1(\cos^2\alpha_1 - \sin^2\theta)]/2 \sin \theta [\sin \theta + \cos \alpha_1 \sin (\alpha_2 - \theta)]$$
(5.64)

Example 5.21 Two straights AB and CD are to be joined by a circular reverse curve with an initial radius of 200 m, commencing at B. From the coordinates given overleaf, calculate the

radius of the second curve which joins the first and terminates at
C (Fig. 5.33).

	E (m)	N (m)
A	103.61	204.82
B	248.86	422.62
C	866.34	406.61
D	801.63	141.88

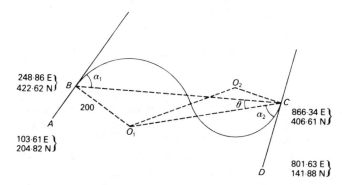

Fig. 5.33

Bearing $AB = \tan^{-1}(248.86 - 103.61)/(422.62 - 204.82)$
$= 33.699°$

Bearing $DC = \tan^{-1}(866.34 - 801.63)/(406.61 - 141.88)$
$= 13.736°$

Bearing $BC = \tan^{-1}(866.34 - 248.86)/(406.61 - 422.62)$
$= 91.485°$

Length $BC = 617.48/\sin 91.485 = 617.69$ m $\quad (T_1T_2)$
$\alpha_1 = 91.485° - 33.699° = 57.786°$
$\alpha_2 = 91.485° - 13.736° = 77.749°$

By equation 5.63,

$\theta = \tan^{-1}(R_1 \cos \alpha_1)/(T_1T_2 - R_1 \sin \alpha_1)$
$= \tan^{-1}(200.00 \cos 57.786)/(617.69 - 200.00 \sin 57.786)$
$= 13.373°$

By equation 5.64,

$R_2 = \dfrac{200.00(\cos^2 57.786 - \sin^2 13.373)}{2 \sin 13.373[\sin 13.373 + \cos 57.786 \sin (77.749 - 13.373)]}$
$= 140.0$ m

Exercises 5.5

1 Two roadways AB and CD are to be connected by a reverse curve of common radius, commencing at B and C. The coordinates of the stations are as follows.

	E (m)	N (m)
A	21 642.87	37 160.36
B	21 672.84	37 241.62
C	21 951.63	37 350.44

If the bearing of the roadway CD is $020° 14' 41''$, calculate the radius of the curve. (Ans. 100.0 m)

2 Two straight railway tracks 300 m apart between centre lines and bearing $012° 00'$ are to be connected by a reverse curve starting from the tangent point A on the centre line of the westerly track and turning in a north-easterly direction to join the easterly track at the tangent point C. The first curve AB has a radius of 400 m and the second BC has a radius of 270 m. The tangent point common to both curves is at B. Calculate (a) the coordinates of B and C relative to the zero origin at A, and (b) the lengths of the curves AB and BC.
(Ans. (a) 244.53 m E, 288.99 m N; 409.58 m E, 484.05 m N; (b) 394.30 m; 266.15 m)

3 Two straights AB and CD are to be joined by a circular reverse curve with an initial radius of 180 m commencing at B. From the coordinates given below, calculate the radius of the second curve which joins the first and terminates at C.

	E (m)	N (m)
A	93.249	184.338
B	223.974	380.358
C	779.706	365.949
D	721.467	127.692

(RICS/M Ans. 634.5 m)

Bibliography

SHEPHERD, F. A., *Surveying Problems and Solutions*, Edward Arnold (1968)

6
Transition curves

A vehicle, of mass M (kg), travelling at a speed v (ms^{-1}), around a circular arc of radius R(m), is subjected to a radial acceleration (v^2/R) producing a centrifugal force F. As F is dependent upon speed, and the radius of curvature is nil along the straight and maximum along the circular arc and thus at the tangent points, its effect is immediate. To overcome this, transition curves are introduced in which the radius is variable, i.e. $r = \infty$ on the straight to $r = R$, the radius of minimum curvature, i.e. the radius of the circular arc. As a circular arc is not always used the minimum radius $r_{min} = R$ will occur at a point. The horizontal component of the transition is thus a spiral, whilst the vertical alignment is dependent upon superelevation (θ), or cant (c), which is introduced to reduce the sideslip generated by the centrifugal force.

6.1 Superelevation (θ)

Considering the factors above, the centrifugal force $F = Mv^2/R$, must be resisted by either the rails, in the case of a railway train, or the adhesion between the road and the vehicles' tyres, unless superelevation is applied, when forces along the plane are equalised (see Fig. 6.1). Then

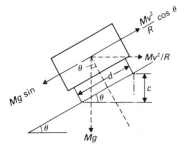

Fig. 6.1

$$Mv^2 \cos\theta/R = Mg\sin\theta \qquad (6.1)$$
$$\tan\theta = v^2/(gR) \qquad (6.2)$$

If θ is small, then

$$\theta_{rad} \simeq v^2/(gR) \qquad (6.3)$$

This is known as the **centripetal ratio**.

6.2.1 Cant (c)

If d is the width of the track, then the cant is given by

$$c = d\sin\theta \qquad (6.4)$$
$$= dv^2 \cos\theta/gR \qquad (6.5)$$
$$\simeq dv^2/gR \qquad (6.6)$$

Note that if θ is small $c \propto v^2 \propto 1/R$.

6.2 Superelevation on roadways

Standards of superelevation tend to be arbitrary but the Ministry of Transport specify the following.

40% of the centrifugal force should normally be balanced out, i.e.

$$\tan\theta = 0.4v^2/gR \quad (v \text{ in m s}^{-2}) \tag{6.7}$$
$$= 0.4V^2/(3.6)^2 \times 9.81R \tag{6.8}$$
$$= V^2/317.85R \tag{6.9}$$

where V is in km h^{-1}.

The following points should be noted.

1) The official formula is given as the gradient 1 in $314R/V^2$ or $V^2/3.14R$ %.
2) The crossfall should not be steeper than 1 in 14.5, i.e. 7%.
3) The superelevation should be such that the residual sideways force, being balanced by the road friction, does not normally exceed $0.15g$, i.e. 1.47 m s^{-2}. (Up to 50 km h^{-1} this may be increased to $0.18g$, i.e. 1.77 m s^{-2}.)

If the superelevation is expressed as 1 in N, then

$V^2/127R$ must not exceed $1/N + 0.15$.

6.3 Minimum curvature for standard velocity

Without superelevation on roads side slip will occur if the side thrust is greater than the adhesion, i.e. if

$Mv^2/R > \mu Mg$

where μ is the coefficient of adhesion, usually = 0.25. Thus the limiting radius is given by

$$R_{\text{min}} = v^2/\mu g \tag{6.10}$$

If the velocity V is given in km h^{-1} then

$$R_{\text{min}} = V^2/3.6^2 g \tag{6.11}$$
$$= V^2/127.14 \tag{6.12}$$

(See MOT recommendation.)

6.4 Length of the transition (L)

The following criteria are suggested

1) An arbitrary length of say 50 m.
2) The total length of the curve is divided into three equal parts: $\frac{1}{3}$ each transition and $\frac{1}{3}$ circular arc.

3) An arbitrary gradient, 1 in x, e.g. 1 in 300, the steepest gradient recommended for railways.
4) At a limited rate of change of radial acceleration. Short in 1908 suggested 1 ft s^{-3} as a suitable value for passenger comfort. In the Highway Transition Curve Tables values of 0.30, 0.45 and 0.60 m s^{-3} are used and this method is considered to be the most suitable criteria for roadway design.

6.5 Radial acceleration

The radial acceleration increases from zero, at the start of the transition, to v^2/R, at the join with the circular curve. The time taken to travel along the transition curve is

$$t = L/v \quad \text{seconds}$$

The rate of gain of radial acceleration is then

$$a = v^2/R \div L/v$$
$$= v^3/RL \quad (\text{m s}^{-3}) \tag{6.13}$$

The length of the curve is then given as

$$L = v^3/aR \tag{6.14}$$
$$= V^3/3.6^3 \, aR \tag{6.15}$$
$$= V^3/46.66 \, aR \tag{6.16}$$

The table given in the 'County Surveyors' Society Treatise' is based upon

$$a = (D/L) \times (V^3/267\,400) \tag{6.17}$$

where D is the degree of the curve (see p. 232). In these tables (D/L) is given as a constant for each page in the tables together with the constant (RL). As

$$D_{\text{rad}} = 100/R$$

(see p. 184, Circular Curves, equations 5.2 and 5.3)

$$D° = 5729.58/R$$
$$R/L = 5729.58/(D/L) \tag{6.18}$$
$$a = V^3/46.66 \, RL = V^3/46.66 \times (D/L)/5729.58$$
$$= V^3 (D/L)/267\,400$$

as (6.17) above.

6.6 The ideal transition curve

If the centrifugal force $F = Mv^2/r$ is to increase at a constant rate it must vary with time and therefore, if the speed is constant,

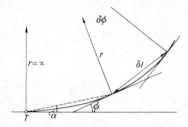

Fig. 6.2 The ideal transition curve

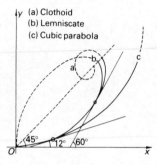

(a) Clothoid
(b) Lemniscate
(c) Cubic parabola

Fig. 6.3

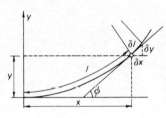

Fig. 6.4 Clothoid

with distance.

$$F \propto l \propto Mv^2/r$$
$$l \propto 1/r$$

i.e. $rl = RL = k$

where k is a constant, R is the radius of the circular arc, and L is the total length of the transition.

In Fig. 6.2,

$$\delta l = r\,\delta\phi$$
$$d\phi = (1/r)\,dl$$
$$= (l/k)\,dl$$

Integrating gives

$$\phi = l^2/2k + c$$

but when $\phi = 0$, $l = 0$ and thus $c = 0$

$$\phi = l^2/2RL$$

This is an intrinsic equation of the clothoid (Fig. 6.3), to which the lemniscate and the cubic parabola are approximations often adopted when the deviation angle is small.

6.7 The clothoid

For the clothoid (Fig. 6.4),

$$\phi = l^2/2RL \tag{6.19}$$

where l = the length of the arc up to a given point, of radius at that point, r

L = the total length of the transition

R = the minimum radius of the transition

$RL = rl = k$, a constant for the curve

As the tangent angle ϕ at any given point is difficult to define, the curve cannot be set out in this intrinsic form.

6.7.1 To find the cartesian coordinates

$$dx/dl = \cos\phi = 1 - \phi^2/2! + \phi^4/4! \quad \text{etc.} \tag{6.20}$$
$$= 1 - l^4/2!(2RL)^2 + l^8/4!(2RL)^4 \quad \text{etc.} \tag{6.21}$$

Integrating (when $x = 0$, $l = 0$),

$$x = l[1 - l^4/(5 \times 2!(2RL)^2)$$
$$+ l^8/(9 \times 4!(2RL)^4) \quad \text{etc.} \ldots] \tag{6.22}$$

For maximum values, i.e. when $x = X$ and $l = L$, then for ϕ_{max}

$$X = L(1 - L^2/40R^2 + L^4/3456R^4 \quad \text{etc.} \ldots) \tag{6.23}$$

Similarly,

$$dy/dl = \sin\phi = \phi - \phi^3/3! + \phi^5/5! \quad \text{etc.} \quad (6.24)$$
$$= l^2/(2RL) - l^6/(3!(2RL)^3)$$
$$+ l^{10}/(5!(2RL)^5) \quad \text{etc.} \quad (6.25)$$

Integrating (when $x = 0$, $l = 0$),

$$y = l[l^2/3(2RL) - l^6/(7 \times 3!(2RL)^3$$
$$+ l^{10}/(11 \times 5!(2RL)^5 \quad \text{etc.}] \quad (6.26)$$

For maximum values

$$Y = L^2/6R[1 - L^2/56R^2 + L^4/7040R^4 + \ldots] \quad (6.27)$$

6.7.2 The tangential angle (α)

$$\tan\alpha = y/x = \frac{l(\phi/3 - \phi^3/42 + \phi^5/1320 + \ldots)}{(1 - \phi^2/10 + \phi^4/216 + \ldots)} \quad (6.28)$$

Expanding binomially gives

$$\tan\alpha \simeq \phi/3 + \phi^3/105 + 26\phi^5/155\,925 + \ldots \text{ etc.} \quad (6.29)$$
$$\simeq \phi/3 + \phi^3/105 + \phi^5/5997 + \ldots \text{ etc.} \quad (6.30)$$

As $\quad \alpha = \tan\alpha - (\tan^3\alpha)/3 + (\tan^5\alpha)/5 + \ldots \text{ etc.} \quad (6.31)$

Substituting into equation above gives

$$\alpha = \phi/3 - 8\phi^3/2835 - 32\phi^5/467\,775 - \ldots \text{ etc.} \quad (6.32)$$
$$= \phi/3 - [8 \times 27(\phi/3)^3/2835]$$
$$- [32 \times 243(\phi/3)^5/467\,775] \quad (6.33)$$

Writing α in degrees

$$\alpha° = \phi°/3 - [(0.0762(\phi°/3)^3 + 0.0166(\phi°/3)^5)]/57.2958 \quad (6.34)$$
$$= \phi°/3 - 3.096 \times 10^{-3}(\phi°/3) \quad \text{secs} \quad (6.35)$$
$$= \phi°/3 - K'' \quad (6.36)$$

For $\quad \phi = 21°$ and $K = 28''$ then $\alpha = 7° - 28''$
$\quad\quad\phi = 24°$ and $K = 43''$ $\quad\quad \alpha = 8° - 43''$
$\quad\quad\phi = 30°$ and $K = 84''$ $\quad\quad \alpha = 10° - 84''$

If $\phi < 20°$ and $K < 20''$ whilst if $\phi < 6$ then K is negligible.

6.7.3 Amount of shift

Shift (s) is the displacement of the circular arc from the tangent, i.e. DF in Fig. 6.5. By equation 6.27,

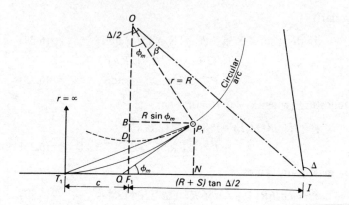

Fig. 6.5

$$PN = BF = y_{\max}$$
$$= L(\phi_m/3 - \phi_m^3/42 + \phi_m^5/1320) \quad \text{etc.} \quad (6.37)$$
$$= L^2/6R - L^4/336R^3 + L^6/42\,240R^5 - \ldots \quad (6.38)$$

The shift is given by
$$s = DF = BF - BD$$
$$= y_{\max} - R(1 - \cos \phi_m)$$

but
$$R(1 - \cos \phi_m) = R[1 - (1 - \phi^2/2 + \phi^4/24 - \phi^6/720)] \quad \text{etc.}$$
$$= L^2/8R - L^4/384R^3 + L^6/46\,080R^5 \quad \text{etc.}$$

then
$$s = (L^2/6R - L^4/336R^3 + L^6/42\,240R^5 - \ldots)$$
$$\quad - (L^2/8R - L^4/384R^3 + L^6/46\,080R^5 - \ldots)$$
$$\simeq L^2/R(1/6 - 1/8) + L^4/R^3(-1/336 + 1/384) + L^6/R^5$$

Thus
$$s \simeq L^2/24R - L^4/2688R^3 + L^6/506\,880R^5 \quad \text{etc.}$$
$$(6.39)$$

For most practical purposes, $s \simeq L^2/24R$. $\quad (6.40)$

6.7.4 Tangent length

The tangent length (see Fig. 6.5) is given by
$$T_1 I = T_2 I$$
$$= (R + s)\tan \Delta/2 + T_1 F_1$$
$$= (R + s)\tan \Delta/2 + (X - R\sin \phi_m)$$
$$= (R + s)\tan \Delta/2 + C \quad (6.41)$$

From equation 6.23,

$$X = L[1 - L^4/[5 \times 2!(2RL)]^2 + L^8/[9 \times 4!(2RL)]^4] \quad \text{etc.}$$
$$= L(1 - L^2/40R^2 + L^4/3456R^4) \quad \text{etc.}$$
$$R \sin \phi_m = L/2 - L^3/48R^2 + L^5/3840R^4 \quad \text{etc.} \quad (6.42)$$

Then

$$C = X - R \sin \phi_m = L(1 - 1/2) - L^3/R^2(1/40 - 1/48)$$
$$+ L^5/R^4(1/3456 - 1/3840) - \ldots$$
$$C = L/2 - L^3/240R^2 + L^5/345\,60R^4 - \ldots \quad \text{etc.} \quad (6.43)$$

The tangent length is given by

$$(R+s)\tan \Delta/2 + L/2 - (L^3/240R^2 - L^5/345\,60R^4 + \ldots)$$
$$(6.44)$$
$$\simeq (R+s)\tan \Delta/2 + L/2 \quad (6.45)$$

6.8 The Bernoulli lemniscate

The polar equation is of the form $c^2 = a^2 \sin 2\alpha$. It is identical to the clothoid for deviation angles up to 60° (the radius decreases up to 135°); c is a maximum when $\alpha = 45$. If the lemniscate approximates to the circle then the polar equation becomes

$$c = 3Rc \sin 2\alpha$$

This is no longer of significance since the advent of the computer.

6.9 The cubic parabola

In the past this was the type of curve most widely used in practice because of its simplicity. It is identical with the clothoid and lemniscate for deviation angles up to 12°. The radius of curvature reaches a minimum for deviation angles of 24° 06′ and then increases. It is therefore not acceptable beyond this point. Let

$$y = x^3/k$$
$$dy/dx = 3x^2/k \quad \text{and} \quad d^2y/dx^2 = 6x/k$$

By the calculus, the curvature (ρ) is given as

$$\rho = 1/r = d^2y/dx^2/[(1 + dy/dx)^2]^{3/2}$$

If ϕ is small, dy/dx is small and $(dy/dx)^2$ may be neglected:

$$1/r = d^2y/dx^2 = 6x/k$$

therefore $\quad k = 6rx = 6RX$

at the end of the curve.

The equation of the cubic parabola is then

$$y = x^3/6RX$$

and the equation of the cubic spiral is

$$y = l^3/6RL$$

when $x \simeq l$, $X \simeq L$.

Note that these approximate to the first term in the clothoid series.

$$\phi \simeq \tan\phi = dy/dx = x^2/2RX$$
$$\alpha \simeq \tan\alpha = y/x = x^2/6RX$$

and $\quad \alpha = \phi/3$

as in the first term of the clothoid.

In Fig. 6.5,

$$PB = R\sin\phi_m \simeq R\phi_m = X/2$$

i.e. the shift bisects the length.

$$DB = R(1 - \cos\phi)$$

The shift (s) is

$$DF = y - R(1 - \cos\phi)$$
$$= X^3/6RX - 2R\sin^2(\phi/2)$$
$$= X^2/6R - 2R\phi^2/4$$
$$= X^2/6R - X^2/8R$$

Then as the first term in the clothoid $s = X^2/24R \simeq L^2/24R$.

6.10 Transition curve tables

Much of the early standard work on transition curves was written by Criswell in *Highway Spirals, Superelevation and Vertical Curves* which includes a set of tables based upon Imperial units. Since the advent of metrication, tables have been prepared by the County Surveyors' Society as *Highway Transition Curve Tables* (*metric*), a treatise on theory and use. These are based upon the clothoid, $\phi = l^2/2RL$ (see Figs 6.6 and 6.7), and tables are given in which the spiral selection is dependent upon the design speed, related to the rate of gain of radial acceleration, together with the constants (D/L), the increase in the degree of curvature per metre and (RL), the product of the radius by the length of spiral.

6.10 TRANSITION CURVE TABLES

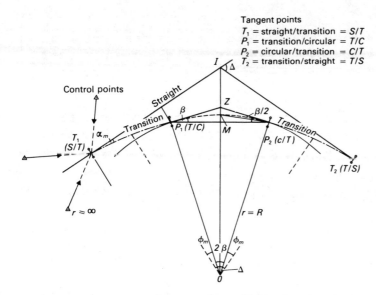

Fig. 6.6 Conventional signs for drawing transition curves

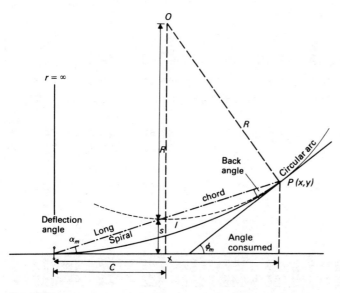

Fig. 6.7

232 TRANSITION CURVES

Table 9 County Surveyors' Society *Highway Transition Curve Tables*

Gain of accn m s⁻³   0.30   0.45   0.60   Increase in degree of curve per metre = $D/L = 0°\ 24'\ 0.0''$
Speed value km h⁻¹   58.5   67.0   73.7 RL constant = 14323.945

Degree of curvature based on 100 m standard arc

Radius R (m)	Degree of curve D ° ′ ″	Spiral length L (m)	Angle consumed ° ′ ″	Shift S (m)	$R+S$ (m)	C (m)	Long chord (m)	Coordinates X (m)	Y (m)	Deflection angle from origin ° ′ ″	Back angle to origin ° ′ ″
2864.7890	2 0 0.0	5.00	0 3 0	0.0004	2864.7893	2.5000	5.0000	5.0000	0.0015	0 1 0.0	0 2 0.0
1432.3945	4 0 0.0	10.00	0 12 0	0.0029	1432.3974	5.0000	10.0000	10.0000	0.0116	0 4 0.0	0 8 0.0
954.9297	6 0 0.0	15.00	0 27 0	0.0098	954.9395	7.5000	15.0000	14.9999	0.0393	0 9 0.0	0 18 0.0
716.1972	8 0 0.0	20.00	0 48 0	0.0233	716.2205	9.9999	19.9998	19.9996	0.0931	0 16 0.0	0 32 0.0
572.9578	10 0 0.0	25.00	1 15 0	0.0455	573.0032	12.4998	24.9995	24.9988	0.1818	0 25 0.0	0 50 0.0
477.4648	12 0 0.0	30.00	1 48 0	0.0785	477.5434	14.9995	29.9987	29.9970	0.3141	0 36 0.0	1 12 0.0
409.2556	14 0 0.0	35.00	2 27 0	0.1247	409.3803	17.4989	34.9972	34.9936	0.4988	0 49 0.0	1 38 0.0
358.0986	16 0 0.0	40.00	3 12 0	0.1861	358.2848	19.9979	39.9945	39.9875	0.7445	1 3 59.9	2 7 59.9
318.3099	18 0 0.0	45.00	4 3 0	0.2650	318.5749	22.4963	44.9900	44.9775	1.0599	1 20 59.8	2 42 59.8
286.4789	20 0 0.0	50.00	5 0 0	0.3635	286.8424	24.9937	49.9831	49.9619	1.4537	1 39 59.6	3 20 59.6
260.4354	22 0 0.0	55.00	6 3 0	0.4838	260.9191	27.4898	54.9727	54.9387	1.9343	2 0 59.3	4 1 59.3
238.7324	24 0 0.0	60.00	7 12 0	0.6280	239.3604	29.9842	59.9579	59.9053	2.5104	2 23 58.8	4 48 58.8
220.3684	26 0 0.0	65.00	8 27 0	0.7982	221.1666	32.4765	64.9372	64.8588	3.1904	2 48 58.1	5 38 58.1
204.6278	28 0 0.0	70.00	9 48 0	0.9967	205.6245	34.9659	69.9090	69.7955	3.9827	3 15 57.1	6 32 57.1
190.9859	30 0 0.0	75.00	11 15 0	1.2255	192.2114	37.4519	74.8716	74.7114	4.8952	3 44 55.6	7 30 55.6
179.0493	32 0 0.0	80.00	12 48 0	1.4867	180.5360	39.9335	79.8227	79.6017	5.9362	4 15 53.5	8 32 53.5
168.5170	34 0 0.0	85.00	14 27 0	1.7824	170.2994	42.4101	84.7600	84.4609	7.1133	4 48 50.6	9 38 9.4
159.1549	36 0 0.0	90.00	16 12 0	2.1145	161.2695	44.8804	89.6806	89.2832	8.4340	5 23 46.8	10 48 13.2
150.7784	38 0 0.0	95.00	18 3 0	2.4852	153.2635	47.3433	94.5816	94.0615	9.9055	6 0 41.8	12 2 18.2
143.2394	40 0 0.0	100.00	20 0 0	2.8963	146.1357	49.7976	99.4595	98.7884	11.5347	6 39 35.2	13 20 24.8
136.4185	42 0 0.0	105.00	22 3 0	3.3496	139.7682	52.2419	104.3105	103.4555	13.3278	7 20 26.7	14 42 33.3
130.2177	44 0 0.0	110.00	24 12 0	3.8471	134.0648	54.6746	109.1303	108.0538	15.2907	8 3 16.0	16 8 44.0
124.5560	46 0 0.0	115.00	26 27 0	4.3905	128.9466	57.0939	113.9144	112.5733	17.4286	8 48 2.4	17 38 57.6
119.3662	48 0 0.0	120.00	28 48 0	4.9814	124.3476	59.4982	118.6579	117.0033	19.7462	9 34 45.6	19 13 14.4
114.5916	50 0 0.0	125.00	31 15 0	5.6214	120.2130	61.8854	123.3551	121.3324	22.2473	10 23 24.9	20 51 35.1
110.1842	52 0 0.0	130.00	33 48 0	6.3120	116.4962	64.2532	128.0004	125.5482	24.9348	11 13 59.5	22 34 0.5
106.1033	54 0 0.0	135.00	36 27 0	7.0544	133.1577	66.5996	132.5873	129.6378	27.8108	12 6 28.7	24 20 31.3
102.3139	56 0 0.0	140.00	39 12 0	7.8499	110.1637	68.9219	137.1091	133.5873	30.8762	13 0 51.5	26 11 8.5
98.7858	58 0 0.0	145.00	42 3 0	8.6994	107.4853	71.2176	141.5585	137.3823	34.1308	13 57 6.9	28 5 53.1
95.4930	60 0 0.0	150.00	45 0 0	9.6040	105.0970	73.4840	145.9278	141.0078	37.5732	14 55 13.7	30 4 46.3
92.4125	62 0 0.0	155.00	48 3 0	10.5642	102.9768	75.7182	150.2090	144.4481	41.2007	15 55 10.8	32 7 49.2
89.5247	64 0 0.0	160.00	51 12 0	11.5807	101.1053	77.9172	154.3933	147.6872	45.0088	16 56 56.5	34 15 3.5
86.8118	66 0 0.0	165.00	54 27 0	12.6536	99.4654	80.0778	158.4717	150.7086	48.8918	18 0 29.3	36 26 30.7
84.2585	68 0 0.0	170.00	57 48 0	13.7830	98.0415	82.1967	162.4347	153.4957	53.1421	19 5 47.4	38 42 12.6
81.8511	70 0 0.0	175.00	61 15 0	14.9687	96.8198	84.2707	166.2722	156.6318	57.4504	20 12 48.7	41 11 11.3
229.1831	25 0 0.0	62.50	7 48 45	0.7097	229.8928	31.2306	62.4484	62.3839	2.8369	2 36 13.5	5 12 31.5
190.9859	30 0 0.0	75.00	11 15 0	1.2255	192.2114	37.4519	74.8716	74.7114	4.8952	3 44 55.6	7 30 4.4
163.7022	35 0 0.0	87.50	15 18 45	1.9438	165.6460	43.6460	87.2226	86.8771	7.7552	5 6 3.9	10 12 41.1
143.2394	40 0 0.0	100.00	20 0 0	2.8963	146.1357	49.7976	99.4595	98.7884	11.5347	6 39 35.2	13 20 24.8
127.3240	45 0 0.0	112.50	25 18 45	4.1130	131.4370	55.8860	111.5271	110.3240	16.3375	8 25 24.6	16 53 20.4

6.10 TRANSITION CURVE TABLES

Table 9, p. 233, may be explained as follows.

Column 3 Spiral length (l) is given in 2.00 m intervals.
Column 1 Radius of curvature (r). As the constant $RL = rl$
= 14 323.945 then $r = RL/l$, e.g. when $l = 40.00$
then $r = 14\,323.945/40.00 = 358.0986$.
Column 2 Degree of curve (D) = $D/L \times L$. For example when
$l = 40.00$ then $D = 0°\,40'\,00'' \times 40 = 26°\,40'\,00''$.
Column 4 Angle consumed (ϕ) = $l^2/2RL$ (equation
6.19). For example, when $l = 40.00$ then $\phi = 40^2/2 \times 8594.367$ rad and

$$\phi° = \frac{57.2958 \times 40^2}{2 \times 8594.367} = 5.3333° = 5°\,20'\,00''$$

Column 5 Shift

$$s = \frac{L^2}{24\,R} - \frac{L^4}{2688\,R^3} + \frac{L^6}{506\,880\,R^5} \text{ etc.} \quad \text{(equation 6.39)}$$

e.g. when $l = 40.00$ then $s = 0.1861$.

Column 6 $R + s$ = Column 1 + Column 5

Column 7 $C = \dfrac{L}{2} - \dfrac{L^3}{240\,R^2} + \dfrac{L^5}{34560\,R^4}$ etc. (equation 6.43)

e.g. when $l = 40.00$ then $C = 19.9979$.

Column 8 Long chord of arc M

$$= M - \frac{M^3}{90\,R^2} + \frac{37M^5}{806\,400\,R^4} \text{ etc.}$$

e.g. when $M = l = 40.00$, chord $= 39.9945$.

Column 9 and 10 Coordinates x and y:

$$x = l - \frac{l^5}{40(RL)^2} + \frac{l^9}{3456(RL)^4} \text{ etc.} \quad \text{(equation 6.22)}$$

$$y = \frac{l^3}{6(RL)} - \frac{l^7}{336(RL)^3} + \frac{l^{11}}{42\,240(RL)^5} \text{ etc.}$$
(equation 6.26)

e.g. when $l = 40.00$ then $x = 39.9875$ and $y = 0.7445$.

Column 11 Deflection angle

$$\alpha = \frac{\phi}{3} - \frac{8\phi^3}{2835} - \frac{32\phi^5}{467\,775} \text{ etc.} \quad \text{(equation 6.32)}$$

Column 12 Back angle (θ) = $\phi - \alpha$

e.g. when $\phi = 3°\,12'\,00''$ and $\alpha = 1°\,03'\,59.9''$
then $\theta = 2°\,08'\,00.1'$.

Note that with the extended use of the computer the need for these tables is diminishing.

6.11 Setting-out processes

Setting out may take one of the following two forms (see Fig. 6.8).

a) From the tangent points using either (i) offsets from the tangent, *or* (ii) deflection angles from the tangent.

b) From control points. This involves the computation of the coordinates of points on the curve. The point itself may be established by polar rays using EDM, or by intersection using two theodolites. Note that it is essential, in order to retain through chainage accuracy, that the standard chord (c_s) should be kept to a working minimum but at no time should it exceed the ratio $c_s/R \leqslant 1/20$ (see p. 188).

In either of these methods the location of the tangent points is essential and these points are related to the tangent length from the intersection point I. If the setting out is by deflection angles or offsets from the tangent, then the location of the tangent point is a first priority and if this is by intersection from the control points, then three rays are desirable. Provided that the maximum value of ϕ is less than 12° the cubic parabola/spiral provides an adequate solution and the setting out computations are very much simplified.

(i) *Offsets from the tangent* (Fig. 6.8(b))
From equation 6.37,

$$y = x^3/6RL = kx^3$$

Then $y_n = kx_n^3$

thus if $x_2 = 2x_1$ and $x_3 = 3x_1$ then $y_2 = 8y_1$ and $y_3 = 27y_1$, etc.

(ii) *Deflection (tangential) angles* (Fig. 6.8(c).)

Here $\tan \alpha = y/x = x^2/6RL$ \hfill (6.46)

If α is small, then

$$\alpha'' = 206\,265x^2/6RL \simeq 206\,265l^2/6RL \tag{6.47}$$

$$\alpha° = 57.2958x^2/6RL \simeq 57.2958l^2/6RL \tag{6.48}$$

At the end of the transition, $l = L$ and then

$$\tan \alpha_m = L^2/6RL = L/6R \tag{6.49}$$

$$\alpha_m'' = 206\,265L/6R \tag{6.50}$$

$$\alpha_m° = 57.2958L/6R \tag{6.51}$$

Check on position of P. At the join of the transition to the circular arc,

$$y_{max} = L^2/6R \tag{6.52}$$

$$= \text{shift} \times 4 = 4s \tag{6.53}$$

6.11 SETTING OUT PROCESSES

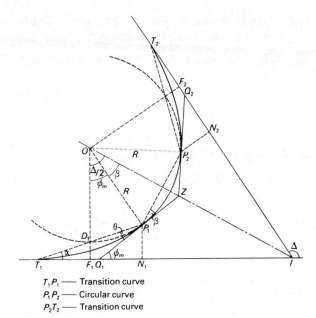

T_1P_1 —— Transition curve
P_1P_2 —— Circular curve
P_2T_2 —— Transition curve

Fig. 6.8(a)

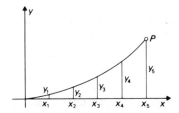

Fig. 6.8(b)

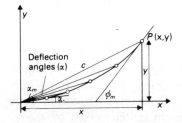

Fig. 6.8(c)

The instrument is now moved to P. Set out the circular curve by offsets or deflection angles from the tangent QPZ. Note that

angle $T_1 P_1 Q_1$ = angle $T_2 P_2 Q_2$ (known as the 'back angle')

$$= \theta = \phi_m - \alpha_m \qquad (6.54)$$
$$\simeq 2\alpha_m \qquad (6.55)$$
$$\tan \phi_m = dy/dx = 3x^2/6RL = x^2/2RL \qquad (6.56)$$
$$\phi''_m = 206\,265 x^2/2RL \simeq 206\,265 L^2/2RL \qquad (6.57)$$
$$\simeq 206\,265 L/2R = 3\alpha''_m \qquad (6.58)$$

Thus $\quad \phi^\circ_m \simeq 57.2958 L/2R \qquad (6.59)$

Consider Fig. 6.8(a). The theodolite is moved from T_1 to P_1, and the following procedure is then adopted.

1) Set the scale of the theodolite to 2α whilst pointing to T_1.
2) Rotate the theodolite until the scale reading is zero. It will then be pointing to Q_1.
3) Transit the telescope (the scale will still be reading zero) and it will now be pointing to Z.
4) Set off the deflection angles to the circular curve until the midpoint (M) on the curve is reached. (Note that the angle MP_1Z should equal $\beta/2$ (where $\beta = \Delta/2 - \phi_m$), and that the angle $P_2 P_1 Z$ should equal β.)
5) Set out the other half of the curve from the other end.

At all times the linear dimensions are set out from peg to peg as in circular curves.

6.12 Application of superelevation on roads

Having decided on the amount of superelevation on the transition curve and thus the maximum amount of cant on the circular curve, a decision must be taken about the manner of applying the cant relative to the centre line. There are three alternatives.

a) Application about the centre line, i.e. an equal amount of cant positive and negative on each side.
b) Application about the inside of the curve.
c) Application about the outside of the curve.

The Institution of Municipal Engineers (Journal, Volume 97, 1970) recommend that the spiral length of a transition curve should be checked to see if it conforms to Roads in Urban Areas, para 3.6. Here the length is determined by the 1% maximum differential gradient requirement. Superelevation is

6.12 APPLICATION OF SUPERELEVATION ON ROADS

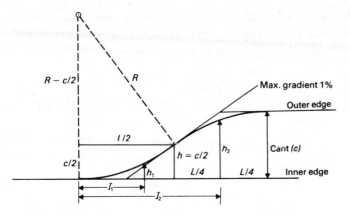

Fig. 6.9

to be applied about the carriageway edge on the inside of the curve by using two reverse vertical curves of equal radius (Fig. 6.9). If $1\% = c/[2(L/4)]$, then

$$L = 200c \tag{6.60}$$

To obtain the levels on the outer kerb,

$$R^2 = (R - c/2)^2 + (L/2)^2$$
$$= R^2 - Rc + c^2/4 + L^2/4$$

Then $R = (c^2 + L^2)/4c \tag{6.61}$

As c is small compared with R, the levels may be computed by the following approximations.

1) For $l < L/2$ $h = l^2/2R = kl^2 \tag{6.62}$
2) For $l > L/2$ $h = c - (L-l)^2/2R$
$$= c - k(L-l)^2 \tag{6.63}$$

Example 6.1 Given $L = 50.000$ m, $c = 0.250$ m, level of $T_1 = 26.320$ m at chainage 1225 m, and standard chord $c_s = 10$ m, then

$$R = (0.250^2 + 50.000^2)/(4 \times 0.250) = 2500.00 \text{ m}$$

Tabulating the values,

Point	Chainage	Length l	Cant c	Level of outer kerb
T_1	1225.00	0	0	26.320
a	1230.00	5.00	0.005	26.325
b	1240.00	15.00	0.045	26.365
c	1250.00	25.00	0.125	26.445
d	1260.00	35.00	0.205	26.525
e	1270.00	45.00	0.245	26.565
P_1	1275.00	50.00	0.250	26.570

Where the compounded transition/circular/transition curve is to be combined with a vertical alignment (Fig. 6.10), the tangent points of the vertical curve should occur as follows:

a) both on the transition curves;
b) both on the circular curve;
c) coincident with the horizontal alignment tangent points P_1 and P_2.

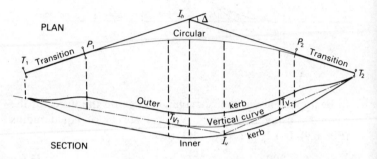

Fig. 6.10

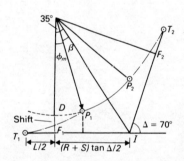

Fig. 6.11

Example 6.2 Two railway lines have straights which are deflected through 70° 00′ 00″ at chainage 832.068. The circular radius is to be 500 m with a maximum cant of 150 mm. The maximum gradient of the outer rail of the transition is to be 1 in 480 (Fig. 6.11). Calculate: (a) the tangent length; (b) the chainages of the tangent points; (c) the setting out data for the transition curve by (i) offsets from the tangent and (ii) deflection angles.

(a) As the maximum gradient is 1 in 480 and the maximum cant is 150 mm then the minimum length of the transition is $L = 480 \times 0.150 = 72.000$ m. The shift is

$$L^2/24R = 72^2/(24 \times 500) = 0.432 \text{ m}$$

The tangent length is

$$T_1 I = (R + s) \tan \Delta/2 + L/2$$
$$= 500.432 \tan 35.0000° + 36.000$$
$$= 386.406 \text{ m}$$

(b) $\phi_m = 57.2958 \, L/2R = 4.1253°$

$\beta = \Delta/2 - \phi_m = 35.0000° - 4.1253°$

$= 30.8747°$

Length of the circular arc is

$$2R\beta_{\text{rad}} = 2 \times 500.000 \times 30.8747/57.2958$$
$$= 538.865 \text{ m}$$

6.12 APPLICATION OF SUPERELEVATION ON ROADS

The chainages are as follows.

$$
\begin{align}
I &= 832.068 \\
TL &= 386.406 \\
T_1 &= 445.662 \\
L &= 72.000 \\
P_1 &= 517.662 \\
\text{arc } P_1P_2 &= 538.865 \\
P_2 &= 1056.527 \\
L &= 72.000 \\
T_2 &= 1128.527
\end{align}
$$

(c) The offsets are based upon

$$y = x^3/6RL = 4.63x^3 \times 10^{-6}$$

The deflection angles are based upon

$$\alpha^\circ = 57.2958\, l^2/6RL = 2.653 l^2 \times 10^{-4}$$

Point	Chainage $x = l$		Offsets	Deflection angles
T_1	445.662			
a	460.000	14.388	0.014	$0.0549 = 0^\circ\, 03'\, 18''$
b	480.000	34.338	0.187	$0.3128 = 0^\circ\, 18'\, 46''$
c	500.000	54.338	0.743	$0.7833 = 0^\circ\, 47'\, 00''$
P_1	517.662	72.000	1.728	$1.3753 = 1^\circ\, 22'\, 31''$

Checks. $y_{max} = 4 \times \text{shift} = 4 \times 0.432 = 1.728$ m

$\alpha_{max} = \phi_{max}/3 = 4.1253/3 = 1.3751 = 1\, 22'\, 30''$

Example 6.3 A circular curve of 600 m radius deflects through an angle of 40° 30' 00". This curve is to be replaced by one of smaller radius so as to admit transitions 100 m long at each end. The deviation of the new curve from the old at their midpoint is 0.500 m towards the intersection point. Determine the amended radius and calculate the length of track to be lifted and the new track to be laid. (RICS/M)
Assume that the change in shift is minimal and is thus based upon the nominated radius of 600 m. The curve is assumed to be of the form $y = x^3/6RL$ (Fig. 6.12). Then

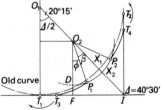

Fig. 6.12

$$\text{shift}(s) = L^2/24R = 100^2/(24 \times 600)$$
$$= 0.694 \text{ m}$$
$$T_1 I = R \tan \Delta/2 = 600 \tan 20.250$$
$$= 221.352 \text{ m}$$
$$O_1 I = R \sec \Delta/2 = 600 \sec 20.250$$
$$= 639.528 \text{ m}$$
$$XI = O_1 I - R = 39.528 \text{ m}$$
$$m = X_1 I = 39.528 - 0.500 = 39.028 \text{ m}$$

Assuming the value of s above, then

$$(R_1 + s)/(R_1 + m) = \cos \Delta/2$$
$$R_1 + s = R_1 \cos \Delta/2 + m \cos \Delta/2$$
$$R_1(1 - \cos \Delta/2) = m \cos \Delta/2 - s$$
$$R_1 = (m \cos \Delta/2 - s)/(1 - \cos \Delta/2)$$
$$= (39.028 \cos 20.250 - 0.694)/(1 - \cos 20.250)$$
$$= 581.17 \text{ m} \quad \text{say} \quad 581 \text{ m}$$

Note that the true value may be obtained by taking the basic equation for shift as

$$s_1 = L^2/24R_1$$

Then $(R_1 + L^2/24R_1)/(R_1 + m) = \cos \Delta/2$

Inserting the values gives

$$24R_1^2 + L^2 = 22.517R_1^2 + 878.777 R_1$$

Then $1.483R_1^2 - 878.777R_1 + 10\,000 = 0$

Solving the quadratic equation gives

$$R_1 = 580.96 \text{ m} \quad \text{say} \quad 581 \text{ m}$$

Then $\phi_{max} = 57.2958 \, L/2R = 4.931$

$$\beta = \Delta/2 - \phi_m \quad = 20.250 - 4.931$$
$$= 15.319 \quad = 0.267_{rad}$$

Then the arc P_1P_2 is given by

$$2R\beta_{rad} = 2 \times 581 \times 0.267 = 310.7 \text{ m}$$

The new track on the curve is

$$2 \times 100 + 310.7 = 510.7 \text{ m} \quad \text{say} \quad 511 \text{ m}$$

The old track on the curve is

$$600 \times 40.5_{rad} = 424.1 \text{ m}$$

The old tangent length is 221.35.
  The new tangent length is

$$(R + s)\tan \Delta/2 + L/2 = 581.69 \tan 20.250 + 50.00$$
$$= 264.60$$

The straights lifted

$$= 2(264.60 - 221.35) = 86.50 \text{ m}$$

The old track lifted

$$= 424.1 + 86.5 = 510.6 \text{ m} \quad \text{say} \quad 511 \text{ m}$$

Example 6.4 The following data is chosen to illustrate the comparison of the cubic spiral with the data shown in the Table 9 of *Highway Transition Curves* (see p. 233) which is based upon the clothoid.

6.12 APPLICATION OF SUPERELEVATION ON ROADS

Increase in degree of curvature per metre = $D/L = 0°24'00''$.
Rate of gain of radial acceleration = 0.30 ms^{-3}.
Radius of curve = 286.4789 m (right handed).
$\Delta = 40°55'05''$ at intersection point I (1000.000 m E, 1000.000 m N, chainage 1000.000 m).
Bearing of straight $T_1 I = 127°48'03''$.
Road width = 7.30 m.

For $D/L = 0°24'00''$, $\alpha = 0.30$ ms^{-3}, and, by equation 6.17,

$$V = (267\,400 \times 0.3/0.4)^{1/3} = 58.53 \text{ kmh}^{-1}$$
$$RL = 5729.58/(D/L) = 14\,323.945$$

Then $L = 14\,323.945/286.4789 = 50.000$ m

Check. $L = v^3/aR$
$$= (58.53/3.6)^3/(0.3 \times 286.4789) = 50.005 \text{ m}$$

This length should now be checked to see if it conforms to equation 6.60 for aesthetic purposes, i.e. $L = 200c$.

$$\text{cant}_{\text{max}} = dv^2/gR \times 0.4$$
$$= (7.30 \times (58.53/3.6)^2 \times 0.4)/(9.81 \times 286.4789)$$
$$= 0.275 \text{ m}$$

Then $L = 200 \times 0.275 = 55.00$ m

It is considered that this is sufficiently close to 50 m to be acceptable.

$$\text{shift} = L^2/24R = 0.364 \text{ m}$$
$$y_{\text{max}} = 4s = 1.456 \text{ m}$$
tangent length $= (R+s)\tan\Delta/2 + C$ (where $C \simeq L/2$)
$= 286.843 \tan 20.4590 + 25.000$
$= 132.012$ m

Coordinates of the tangent points (T_1 and T_2).

Line IT_1: Length = 132.012 bearing $307°48'03''$
$\Delta E = -104.309$ $\Delta N = 80.913$

Then T_1 is 895.691 E, 1080.913 N.
Line IT_2: Length = 132.012, bearing $168°43'08''$
$\Delta E = 25.825$ $\Delta N = -129.461$

Then T_2 is 1025.825 E, 870.539 N.

But $\phi_m = 57.2958\, L/2R = 57.2958 \times 50.000/2 \times 286.4789$
$= 5.0000°$

$\beta = \Delta/2 - \phi_m = 20.4590° - 5.0000°$
$= 15.4590°$

The arc $P_1 P_2$ is given by

$$2R\beta = 2 \times 286.4789 \times (15.4590/57.2958) = 154.590 \text{ m}$$

Chainages

$$I = 1000.000$$
$$TL = 132.012$$
$$T_1 = 867.988$$
$$L = 50.000$$
$$P_1 = 917.988$$
$$\text{arc } P_1P_2 = 154.590$$
$$P_2 = 1072.578$$
$$L = 50.000$$
$$T_2 = 1122.578$$

Setting out data—right hand half of curve ($c_s \leqslant R/20$, say 20 m here).

Using deflection angles

$$\alpha_s = 57.2958l^2/6RL = kl^2 = 0.000\,66l^2$$
$$\alpha_c = 57.2958c/2R = Kc = 0.1c$$

Point	Chainage	l	α	Setting out angle (δ)
Transition				
T_1	867.988			
a	880.000	12.012	0.0962	0° 05′ 46″
b	900.000	32.012	0.6832	0° 40′ 59″
P_1	917.988	50.000	1.6667	1° 40′ 00″
(Check: $\alpha_m = \phi_m/3 = 5.0000/3$)				
Circular arc to midpoint		c		
c	920.000	2.012	0.2012	0° 12′ 04″
d	940.000	20.000	2.0000	2° 12′ 04″
e	960.000	20.000	2.0000	4° 12′ 04″
f	980.000	20.000	2.0000	6° 12′ 04″
M	995.283	15.283	1.5283	7° 43′ 46″
(Check: $\beta/2 = 7° 43′ 46″$)				
g′	1000.000	20.000	2.0000	7° 15′ 28″
f′	1020.000	20.000	2.0000	5° 15′ 28″
e′	1040.000	20.000	2.0000	3° 15′ 28″
d′	1060.000	20.000	2.0000	1° 15′ 28″
Transition		l		
P_2	1072.578	50.000	1.6667	1° 40′ 00″
c′	1080.000	42.578	1.2086	1° 12′ 31″
b′	1100.000	22.578	0.3398	0° 20′ 23″
a′	1120.000	2.578	0.0044	0° 00′ 16″
T_2	1122.578			

To compute the coordinates of P_1, P_2 and M.
This involves the transposition of coordinate technique shown on p. 205, i.e.

6.12 APPLICATION OF SUPERELEVATION ON ROADS

$$E_p = E_{T1} + x\cos\theta - y\sin\theta$$
$$N_p = N_{T1} + y\cos\theta + x\sin\theta$$
$$\theta = \text{angle of swing}$$
$$= \text{old bearing} - \text{new bearing} = 90° - \phi_{T1/I}$$

For P_1,

$$x = 50.000 \text{ m}$$
$$y = x^3/6RL = -50^3/(6 \times 14\,323.945)$$
$$= -1.454$$
$$\theta = 90.0000 - 127.8008 = -37.8008$$

Then $E_{P1} = 895.691 + (50.000 \times 0.790\,146)$
$$- (-1.454 \times -0.612\,918)$$
$$= 895.691 + 39.507 - 0.891 = 934.307 \text{ m}$$
$$N_{P1} = 1080.913 + (-1.454 \times 0.790\,146)$$
$$+ (50.000 \times -0.612\,918)$$
$$= 1080.913 - 1.149 - 30.646 = 1049.118 \text{ m}$$

Bearing of tangent at P_1 is

$$127.8008° + 5.0000° = 132.8008°$$

Angle of swing is

$$\theta = 90.0000° - 132.8008° = -42.8008°$$

For M,

$$l = P_1P_2/2 = 154.590/2$$
$$\psi = 154.590/(2 \times 286.4789) = 0.2698 \text{ rad}$$
$$x = R\sin\psi = 76.361$$
$$y = -R(1-\cos\psi) = -10.364$$

Then

$$E_m = 934.307 + (76.361 \times 0.733\,72) - (-10.364 \times -0.679\,45)$$
$$= 983.293 \text{ m}$$
$$N_m = 1049.118 + (-10.364 \times 0.733\,72) + (76.361 \times -0.679\,45)$$
$$= 989.630 \text{ m}$$

For P_2, the process is similar to the above with

$$\psi = 0.5396 \qquad x = 147.191 \quad \text{and} \quad y = -40.705$$

Then $E_{P2} = 1014.647 \text{ m}$ and $N_{P2} = 919.243 \text{ m}$

Check on the coordinates of M.

$$\text{length } IM = (R+s)\sec\Delta/2 - R = 19.675$$
$$\text{bearing } IM = 127.8008 + 40.9181 + 180.0000$$
$$+ (180.0000 - 40.9181)/2 - 360.0000$$
$$= 058.2599$$

Then $\Delta E_{IM} = -16.732 \quad E_M = 983.268 \text{ m}$
$\Delta N_{IM} = -10.350 \quad N_M = 989.650 \text{ m}$

Comparing these values with those above it is thought that the approximations made in the basic equations produce the discrepancies.

Example 6.5 (Based upon the clothoid, Table 9), with $L = 50.000$ m, $RL = 14\,323.945$, $R = 286.4789$, $C = 24.994$, $s = 0.364$, $\alpha_m = 1°\,39'\,59.6''$, $\phi_m = 5°\,00'\,00''$, $x = 49.962$, $y = -1.454$. Then

$$\text{TL} = (R+s)\tan\Delta/2 + C = 132.007 \quad (\text{cf. }132.012)$$

Coordinates of tangent points.

For T_1,

$$\Delta E_{I/T1} = -104.305$$
$$E_{T1} = 895.695 \text{ m} \quad (\text{cf. }895.691)$$
$$\Delta N_{I/T1} = 80.910$$
$$N_{T1} = 1080.910 \text{ m} \quad (\text{cf. }1080.913)$$

For P_1, there is no change in the values of x and y and thus based upon the above

$$E_{P1} = 934.281 \text{ m} \quad (\text{cf. }934.307)$$
$$N_{P1} = 1049.138 \text{ m} \quad (\text{cf. }1049.118)$$

The arc $P_1 P_2$ does not change here.

For M,

$$E_M = 983.266 \text{ m} \quad (\text{cf. }983.293)(983.268)$$
$$N_M = 989.650 \text{ m} \quad (\text{cf. }989.630)(989.650)$$

For P_2,

$$E_{P2} = 1014.621 \text{ m} \quad (\text{cf. }1014.647)$$
$$N_{P2} = 919.263 \text{ m} \quad (\text{cf. }919.243)$$

Note that the discrepancies arise from

(a) the approximation $X = L$, i.e.

$$X = 49.962 \text{ m} \quad L = 50.000 \text{ m} \quad \Delta = 0.038 \text{ m}$$

(b) the approximation $L/2 = C$, i.e.

$$L/2 = 25.000 \text{ m} \quad C = 24.994 \text{ m} \quad \Delta = 0.006 \text{ m}$$

6.13 Setting out of the transition curve where an obstruction prevents a deflection angle from the tangent point

The following is based upon the equation

$$y = x^3/6RX = l^3/6RL$$

where $x \simeq l$.

6.13 SETTING OUT OF THE TRANSITION CURVE WITH AN OBSTRUCTION

To set out deflection angles from a point B on the transition curve to points A and C (Fig. 6.13), let B be a distance l from T and let the distances BA and BC be of length c. The angle to the tangent at Q and B will be ϕ and let the deflection angle to the transition at B (i.e. CBW) = β, and the tangential angles to C and A respectively be α_1 and α_2. Then

$$\tan \phi \simeq \phi = dy/dx = l^2/2RL$$
$$\tan \alpha \simeq \alpha = y/x = l^2/6RL$$

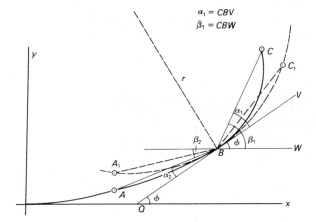

Fig. 6.13

To set out from B to C,

$$\alpha_1 = \beta_1 - \phi$$
$$\tan \beta_1 \simeq \beta_1 = (y_c - y_b)/(x_c - x_b)$$
$$= ((l+c)^3 - l^3)/6RL(l+c-l)$$
$$= (l^3 + 3l^2c + 3lc^2 + c^3 - l^3)/6RLc$$
$$= 3l^2/6RL + 3lc/6RL + c^2/6RL$$
$$= l^2/2RL + lc/2RL + c^2/6RL$$

Then
$$\alpha_1 = l^2/2RL + lc/2RL + c^2/6RL - l^2/2RL$$
$$= lc/2RL + c^2/6RL$$

But as $RL = rl = K$ then $l = K/r$, and

$$\alpha_1 = c/2r + c^2/6K \qquad (6.64)$$
$$= \alpha_c + \alpha_s$$

where α_c is the deflection angle for a chord c to a circular curve of radius r, and α_s is the deflection angle for a chord c to a spiral.
Alternatively the equation may be written as

$$\alpha_1 = c(l + c/3)/2K \qquad (6.65)$$
and $$\alpha_1'' = 206\,265c(l + c/3)/2K \qquad (6.66)$$
$$\alpha_1'' = K'c(l + c/3)$$

where $K' = 206\,265/2RL$. (6.67)

To obtain the deflection angle α_2, by a similar approach,
$$\begin{aligned}\alpha_2 &= \phi - \beta_2 \\ &= l^2/2K - [l^3 - (l-c)^3]/6Kc \\ &= lc/2K - c^2/6K \quad (6.68) \\ &= \alpha_c - \alpha_s\end{aligned}$$

Then $\alpha_2 = c(l - c/3)/2K$ (6.69)

$\alpha_2'' = K'c(l - c/3)$ (6.70)

Example 6.6 Let $L = 60.000$ m, $c = 10.000$ m, $RL = 6000$. At $l = 30.000$ then

$r = 200.000$ and $K'' = 206\,265/(2 \times 6000) = 17.189$

Then $\alpha_1'' = K'c(l + c/3)$
$= (17.189 \times 10.000)(30.000 + 10.000/3)$
$= 5729.6''$ (i.e. $1° 35' 30''$)

and $\alpha_2'' = (17.189 \times 10.000)/(30.000 - 10.000/3)$
$= 4583.7''$ (i.e. $1° 16' 24''$)

The component parts computed individually gives $\alpha_c = 5156.6''$ and $\alpha_s = 573.0''$. Then

$\alpha_1'' = \alpha_c + \alpha_s$
$= 5156.6 + 573.0 = 5729.6''$

$\alpha_2'' = \alpha_c - \alpha_s$
$= 5156.6 - 573.0 = 4583.6''$

6.14 Transition curves connecting compound curves

6.14.1 Curves of the same hand

In this case $R_1 > R_2$ (see Fig. 6.14). The shift between the circular arc and the straight is calculated in the usual way, i.e.

$s_1 = L_1^2/24R_1 - L_1^4/2688R_1^3$ etc.
$\simeq L_1^2/24R_1$

$s_2 = L_2^2/24R_2 - L_2^4/2688R_2^3$ etc.
$\simeq L_2^2/24R_2$

The clearance (s_3) between the two circular curves may be seen as the relative displacement of the centres O_1 and O_2. The movements along the x and y axes are

$\Delta x = C_2 - C_1$

where $C = L/2 - L^3/240R^2 \ldots$ etc.

and $\Delta y = (R_1 + s_1) - (R_2 + s_2)$

6.14 TRANSITION CURVES CONNECTING COMPOUND CURVES

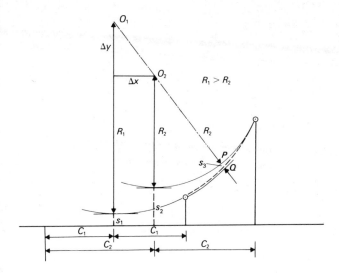

Fig. 6.14

On the line O_1Q,

$$O_1Q = O_1O_2 + O_2P + PQ$$

then $R_1 = (\Delta x^2 + \Delta y^2)^{1/2} + R_2 + s_3$

$$s_3 = \pm[(C_2 - C_1)^2 + (R_1 + s_1 - R_2 - s_2)^2]^{1/2} - (R_2 - R_1) \tag{6.71}$$

This value is approximately equal to the shift appropriate for a transition length $l = L_1 - L_2$. Note that an error will occur in this approximation when ϕ_{max} is large. In Fig. 6.15, if the length of the spiral is given as l, then

$$a \simeq m^3/6RL \tag{6.72}$$
$$b \simeq (l-m)^3/6RL \tag{6.73}$$
$$s_3 = a + b$$
$$= (l^3 - 3l^2m + 3lm^2)/6RL$$

The minimum value of s_3 will occur on the line O_1O_2 produced, i.e.

$$d(s_3)/dm = (-3l^2 + 6lm)/6RL = 0$$
$$m = l/2$$

i.e. the transition is bisected by the clearance s_3 and

$$a = b = s_3/2 = l^3/48RL$$
$$s_3 = l^3/24RL = l^2/24r \tag{6.74}$$

The value RL is assumed constant for all three transitions as

$$L = v^3/aR$$

and $RL = v^3/a = $ constant

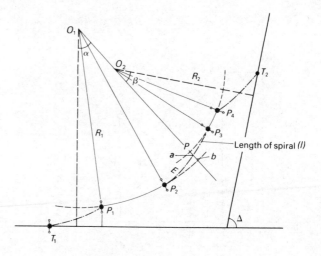

Fig. 6.15

Referring to the compound curve equations (5.48 and 5.49), the tangent lengths may be obtained as follows

1) $t_1 \sin \Delta = (R_2 - R_1)(1 - \cos \beta) + R_1(1 - \cos \Delta)$
2) $t_2 \sin \Delta = (R_1 - R_2)(1 - \cos \alpha) + R_2(1 - \cos \Delta)$ (6.75)

Expanding these gives

1) $t_1 \sin \Delta = R_2 - R_2 \cos \beta - R_1 + R_1 \cos \beta + R_1 - R_1 \cos \Delta$
$= R_2 - R_1 \cos \Delta - (R_2 - R_1) \cos \beta$ (6.76)

and applied to the transition figure,

$t_1 \sin \Delta = (R_2 + s_2) - (R_1 + s_1) \cos \Delta - (R_2 - R_1 + s_3) \cos \beta$

Similarly,

2) $t_2 \sin \Delta = R_1 - R_2 \cos \Delta - (R_1 - R_2) \cos \alpha$

becomes $(R_1 + s_1) - (R_2 + s_2) \cos \Delta - (R_1 - R_2 - s_3) \cos \alpha$

(6.77)

To compute the through chainages,

The circular arc $P_1 P_2 = R \alpha_{\text{rad}} - L/2 - l/2$ (6.78)
The circular arc $P_3 P_4 = R \beta_{\text{rad}} - L/2 - l/2$ (6.79)

The total arc length is then given as

$T_1 P_1 + P_1 P_2 + P_2 P_3 + P_3 P_4 + P_4 T_2$

i.e. $L_1 + (R\alpha - L/2 - l/2) + l + (R\beta - L/2 - l/2) + L_2$ (6.80)

6.14.2 Curves of the opposite hand

With reference to Fig. 6.16, let $P_2Q = m$ and let $QP_3 = l - m$. Then Δx and Δy represent the movement of the centres of the circular arcs in the x and y directions to allow the transition curve P_2P_3 to be inserted, and

$$\Delta x = C_1 + C_2$$
$$\Delta y = (R_1 + s_1) + (R_2 + s_2)$$

The minimum clearance lies on the line O_1O_2:

$$O_1O_2 = R_1 + s_3 + R_2$$
$$= (\Delta x^2 + \Delta y^2)^{1/2}$$
$$s_3 = (\Delta x^2 + \Delta y^2)^{1/2} - (R_1 + R_2)$$
$$= [(C_1 + C_2)^2 + (R_1 + s_1 + R_2 + s_2)^2]^{1/2} - (R_1 + R_2) \quad (6.81)$$

Given $a \simeq m^3/6RL$ (equation 6.104),

$$b \simeq (l-m)^3/6RL$$
$$s_3 \simeq a + b$$
$$= (m^3 + l^3 - 3l^2m + 3lm^2 - m^3)/6RL$$

For the minimum value,

$$d(s_3)/dm = -3l^2 + 6lm = 0$$
$$m = l/2$$

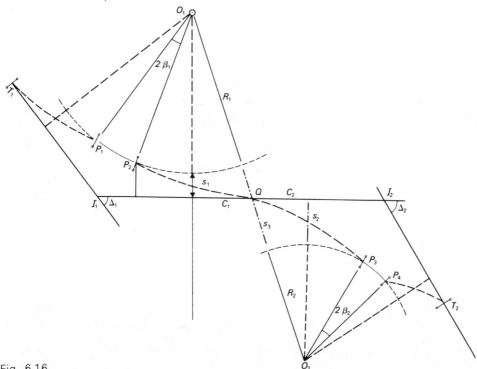

Fig. 6.16

Then $a = b = s_3/2 = l^3/48RL$

$$s_3 = l^3/24RL = l^2/24r \qquad (6.82)$$

To compute the through chainage,

$$\begin{aligned}
T_1T_2 &= T_1P_1 + P_1P_2 + P_2P_3 + P_3P_4 + P_4T_2 \\
&= L_1 + 2R\beta_1 + l + 2R\beta_2 + L_2 \qquad (6.83) \\
&= L_1 + (R\Delta_1 - L_1/2 - l/2) + l \\
&\quad + (R\Delta_2 - L_2/2 - l/2) + L_2 \qquad (6.84)
\end{aligned}$$

The tangent lengths are computed as follows:

$$T_1I_1 = (R_1 + s_1)\tan\Delta_1/2 + C_1 = T_1Q \qquad (6.85)$$
$$T_2I_2 = (R_2 + s_2)\tan\Delta_2/2 + C_2 + T_2Q \qquad (6.86)$$

Example 6.7 Transition curve for a compound curve with radii of the same hand. Given $\Delta = 75°00'00''$, $\alpha = 30°00'00''$, $R_1 = 250.000$ m, $R_2 = 300.000$ m, $V = 80$ km h^{-1}, $a = 0.3$ m s^{-3}, chainage of $I = 872.700$ m. Then

$$L_1 = (80/3.6)^3/(0.3 \times 250.000) = 146.319 \text{ m}$$
$$RL = 146.319 \times 250.000 = 36\,579.790$$

and $L_2 = 36\,579.790/300.000 = 121.933$ m

$$\begin{aligned}
s_1 &= L_1^2/24R_1 - L_1^4/2688R_1^3 \\
&= 3.568 - 0.011 = 3.557 \text{ m}
\end{aligned}$$

Similarly,

$$s_2 = 2.065 - 0.003 = 2.062 \text{ m}$$

Then $l = L_1 - L_2 = 24.386$ m

$$s_3 = l^3/24RL = 0.017 \text{ m}$$
$$\begin{aligned}
C_1 &= L_1/2 - L_1^3/240R_1^2 \ldots \\
&= 73.159 - 0.209 = 72.950 \text{ m}
\end{aligned}$$

Similarly,

$$C_2 = 60.967 - 0.084 = 60.883 \text{ m}$$

Check. $s_3 = [(R_1 + s_1 - R_2 - s_2)^2 + (C_1 - C_2)^2]^{1/2} - (R_1 - R_2)$ (note that $R_1 < R_2$)

$$= \pm 49.983 - 50.000 = 0.017 \text{ m}$$

But $R_1 + s_1 = 253.557$ m and $R_2 + s_2 = 302.062$ m

Then $O_1O_2 = R_2 - R_1 - s_3$

$$= 300.000 - 250.000 - 0.017 = 49.983 \text{ m}$$

$$T_1I = t_1 + C_1 \quad \text{and} \quad T_2I = t_2 + C_2$$

$t_1 \sin\Delta = (R_2 + s_2) - (R_1 + s_1)\cos\Delta + (R_2 - R_1 - s_3)\cos\beta$

$$t_1 = (302.062 - 253.557 \cos 75 + 49.983 \cos 45)/\sin 75$$
$$= 281.367 \text{ m}$$
$$t_2 \sin \Delta = (R_1 + s_1) - (R_2 + s_2) \cos \Delta - (R_2 - R_1 - s_3) \cos \alpha$$
$$t_2 = (253.557 - 302.062 \cos 75 - 49.983 \cos 30)/\sin 75$$
$$= 136.751 \text{ m}$$

Exercises 6.1

1 A road curve of 600 m radius is to be connected to two straights by means of transition curves of the cubic parabola type at each end. The maximum speed on this part of the curve is to be 100 km h^{-1} and the rate of change of radial acceleration is 0.3 m s^{-3}. The angle of intersection is 50° 00′ 00″ and the chainage of the intersection point is 1790.04 m. Calculate: (a) the length of each transition curve; (b) the shift of the circular arc; (c) the chainage at the beginning and end of the composite curve; (d) the value of the first two deflection angles for setting out the first two pegs of the transition curve from the first tangent point, assuming the two pegs are set out at 20 m, based upon through chainage.

(ICE Ans. (a) 119.08 m; (b) 0.985 m; (c) 1450.26, 2092.94 m; (d) 46″, 07′ 06″)

2 Two tangents which intersect at an angle of 41° 40′ 00″ are to be connected by a circular curve of 1000 m radius with a transition curve at each end. The chainage of the intersection point is 8486.42 m. The transition curves are to be of the cubic parabolic type, designed for a maximum speed of 80 km h^{-1}, and the rate of change of radial acceleration is not to exceed 0.3 m s^{-3}. Find the chainage of the beginning and end of the first transition curve and draw up a table of deflection angles for setting out the curve in 20 m chord lengths chainage running continuously through from the tangent point.

(ICE Ans. 8087.58, 8124.16; 02′ 25″, 16′ 28″, 20′ 58″)

3 The limiting speed around a circular curve of 600 m radius calls for a superelevation of 1/24 across the 10.0 m carriageway. Adopting the Ministry of Transport's recommendation that the length of transition should approximate to the length determined from the 1 % differential gradient requirements (RUA para 3.6) where superelevation is to be applied about the carriageway edge on the inside of the curve by using two reverse vertical curves of equal radius, calculate: (a) the tangential angles for setting out the transition curve at 20 m intervals from the tangent point; (b) the levels of the outer edge of the carriageway at 20 m intervals based upon the above recommendations.

(TP Ans. 04' 35", 18' 20", 41' 15", 1° 13' 20", 1° 19' 35"; 0.048 m, 0.192 m, 0.352 m, 0.416 m, 0.417 m)

4 Two straights of a proposed length of railway track are to be joined by a circular curve of 700 m radius with cubic parabolic transitions 70 m long at entry and exit. The deflection angle between the two straights is 22° 38' 00" and the chainage of the intersection point on the first straight produced is 2553.00 m. Determine the chainages of the ends of both transitions and the information required for the setting out of the midpoint and the end of the first transition. If the transition curve is designed to give a rate of change of radial acceleration of 0.3 m s^{-3}, what will be the superelevation of the outer rail at the midpoint of the transition, if the distance between the rails is 1.500 m.
(ICE Ans. 2377.86, 2447.86, 2654.38, 2724.38; shift 0.292 m, TL 175.144 m, 0° 14' 19", 0° 57' 18"; 0.066 m)

5 Two straight portions of a railway line, intersecting at an angle of 155° are to be connected by two cubic parabolic curves, each 100 m long, and a circular curve of 400 m radius. Calculate the necessary data for setting out the curve using chords 20 m long.
(RICS/M Ans. 40 000 RL; 05' 44", 22' 55", 51' 34", 1° 31' 40", 2° 23' 14", shift 1.042 m, TL 1858.98 m)

6 Two straights of a railway with 1.435 m gauge intersect at an angle of 135°. They are to be connected by a curve of 240 m radius with cubic parabolic transition at either end. The curve is to be designed for a maximum speed of 50 km h^{-1} with a rate of change of radial acceleration of 0.3 m s^{-3}. Calculate: (a) the required length of transition; (b) the maximum superelevation of the outer rail; (c) the amount of shift required for the transition; and (d) the lengths of the tangent points from the intersection of the straights.
(RICS/M Ans. 37.21; 0.118; 0.240; 598.597 m)

7 (a) What are the conditions which should be fulfilled by an 'ideal' transition curve?
(b) Why are cubic parabolae and cubic spirals acceptable in practice and what are their limiting conditions?
(c) Describe how you would set out a complete curve layout consisting of a circular arc with terminal transitions, emphasising the methods of checking the individual stages. (RICS/M)

8 The curve connecting two straights is to be wholly transitional without intermediate circular arc, and the junction of the two transitional curves is 5.000 m from the intersection point of the straights which deflects through 18° 00' 00". Calculate the tangent distances and the minimum radius of curvature.
If the superelevation is limited to 1 in 16 determine the correct velocity for the curve and the rate of gain of radial acceleration.
(LU Ans. 94.89 m, 300.88 m, 49 km h^{-1}, 0.09 m s^{-3})

9 As part of a highway realignment, it is required to join two circular arcs with a cubic spiral transition. The design speed is 90 km h^{-1} with a rate of gain of radial acceleration of 0.3 m s^{-3}. The two existing radii are 300 m and 500 m which subtend 60° and 45° respectively. Calculate the length of the transition curves, shifts and the setting out data based upon the osculating circle theory, where necessary, for the standard chords of 20 m of 'through chainage' from the first tangent point on the straight, which has a chainage of 12 + 34.620 m.
(Ans. T_1 1234.620, P_1 1408.231, P_2 1600.863, P_3 1670.307, P_4 1976.200 T_2 2080.367)

Bibliography

ALLAN, A. L., HOLLWEY, J. R., and MAYNES, J. H. B., *Practical Field Surveying and Computations.* Heinemann (1973)

COUNTY SURVEYORS' SOCIETY, *Highway Transition Curve Tables—A Treatise on Theory and Use.* Carriers Publishing Co. Ltd (1969)

HMSO, *Layout of Roads in Rural Areas and Technical Memorandum.* Department of the Environment (1968)

INSTITUTION OF MUNICIPAL ENGINEERING, *Journal*, 97 (1970)

JENKINS, R. B. M., *Curve Surveying.* Macmillan

ROYAL DAWSON, F. G., *Elements of Curve Design for Road, Railway and Racing Track.* Spon (1932)

ROYAL DAWSON, F. G., *Road Curves for Safe Modern Traffic.* Spon (1932)

SHEPHERD, F. A., *Surveying Problems and Solutions.* Edward Arnold (1968)

SHORTT, W. H., A practical method for improvement of existing railway curves, *Proc. Inst. Civil Eng.*, 176

7
Vertical curves

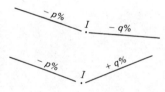

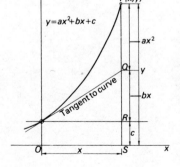

Fig. 7.1

Where it is required to smooth out a change of gradient some form of parabolic curve is used. There are two general forms: **convex** (or summit) curves, and **concave** (sag or valley) curves (Fig. 7.1). The properties required are as follows.

a) Good riding qualities, i.e. a constant change of gradient and a uniform rate of increase of centrifugal force.
b) Adequate sighting properties.

The simple parabola is normally used because of its simplicity and constant change of gradient, but recently the cubic parabola has come into use, particularly for valley intersections. It has the advantage of a uniform rate of increase of centrifugal force and less filling is required.

Gradients are generally expressed as 1 in x, i.e. 1 vertical to x horizontal, but in vertical curve calculations percentage gradients are mainly used. For example, in the gradient of the first straight is $b \equiv p\%$; the gradient of the second straight is $e \equiv q\%$.

Note that 1 in $x = 100/x\%$, e.g. 1 in 5 = 20%. Gradients rising left to right are considered positive. Gradients falling left to right are considered negative.

7.1 Properties of the simple parabola

$$y = ax^2 + bx + c \quad (7.1)$$
$$dy/dx = 2ax + b$$
$$x = -b/2a \text{ for max. or min.} \quad (7.2)$$
$$d^2y/dx^2 = 2a \quad (7.3)$$

i.e. a constant rate of change of gradient. For a valley curve, a is $+$ ve and for a summit curve, a is $-$ ve. The value of b determines the maximum or minimum position along the x axis. The value of c determines where the curve cuts the y axis.

Fig. 7.2

The difference in elevation between a vertical curve and a tangent to it is equal to half the rate of change of the gradient times the square of the horizontal distance from the point of tangency (Fig. 7.2). As

$$dy/dx = 2ax + b \text{ (grade of the tangent)}$$

When $x = 0$, the grade of the tangent $= b$ and the value of $y = c$.
Then

$$QR = bx \quad \text{and} \quad RS = c$$

and therefore

$$PQ = ax^2$$
$$y = PQ + QR + RS$$
$$= ax^2 + bx + c$$

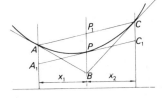

Fig. 7.3

The horizontal lengths of any two tangents from a point to a vertical curve are equal (Fig. 7.3).

$$PB = ax_1^2 = ax_2^2 = AA_1 = CC_1 = PP'$$

Therefore
$$x_1 = x_2$$

The following points should be noted.

i) A, C and P are tangents to the curve, AC is a chord parallel to the tangent A_1C_1 (through P), AA_1, PP' and CC_1 are equal vertical lengths, and thus ACC_1A_1 is a parallelogram.

ii) In vertical curves only horizontal lengths and vertical heights are used.

7.2 Properties of the vertical curve

Compare Fig. 7.4 with Fig. 7.2. Both are based upon the equation $y = ax^2 + bx + c$. Here $PQ = -ax_p^2$, $QR = bx = p\%$ of x and $RS = c =$ the level of T_1 above the datum. The level on the curve at P is then

$$y_p = -ax_p^2 + px_p/100 + y_{T1}$$

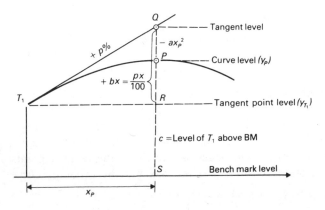

Fig. 7.4

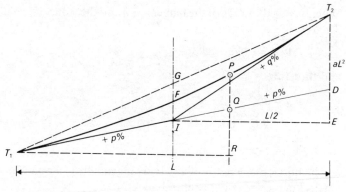

Fig. 7.5

In Fig. 7.5,

$$DT_2 = aL^2 = ET_2 - ED$$

where L is the horizontal length of the curve

$$DT_2 = aL^2 = qL/200 - pL/200 = L(q-p)/200$$

Then
$$a = (q-p)/(200\,L) \tag{7.4}$$

Length of curve

$$L = (q-p)/(200\,a) \tag{7.5}$$

Note that the rate of change of gradient ($2a$) is positive when $p < q$ (valley curves) and negative when $p > q$ (summit curves).

The distance from the intersection point to the curve is given as

$$FI = FG = a(L/2)^2 = aL^2/4$$
$$= (q-p)L^2/800 \tag{7.6}$$

7.2.1 Maximum or minimum height on the curve

By equation 7.2,

$$\begin{aligned}x &= -b/2a \text{ (for max./min.)} \\ &= -p/100/[2 \times (q-p)/200\,L] \\ &= pL/(p-q)\end{aligned} \tag{7.7}$$

Example 7.1 An existing road of rising gradient 1 in 40 meets a falling gradient of 1 in 50, at a chainage 364.370 m and level 50.360 m AOD. The gradients are to be connected by a simple parabolic curve 100 m long. Calculate (a) the rate of change of gradient; (b) the chainages and levels of the tangent points; (c) the chainage and level of the maximum height on the curve.

(a) $\quad p = +100/40 = +2.5\,\%$
$\quad\quad\quad q = -100/50 = -2.0\,\%$

Then $a = (q-p)/200\,L$
$= (-2.0 - 2.5)/20\,000 = -2.25 \times 10^{-4}$
$2a = -4.5 \times 10^{-4}$

i.e. the rate of change of gradient.

(b) Chainage of T_1 = chainage of $I - L/2$
$= 364.370 - 50.000 = 314.370$ m

Chainage of T_2 = chainage of $I + L/2$
$= 364.370 + 50.000 = 414.370$ m

Level of T_1 = level of $I - pL/200$
$= 50.360 - 1.250 = 49.110$ m

Level of T_2 = level of $I - qL/200$
$= 50.360 - 1.000 = 49.360$ m

(c) Chainage of maximum height
$=$ chainage of $T_1 + pL/(p-q)$
$= 314.370 + (2.5 \times 100/4.5) = 369.926$ m

Level of $P = ax^2 + bx + c$
$= (-2.25 \times 10^{-4})(55.556^2)$
$+ (2.5 \times 55.556/100) + 49.110$
$= -0.694 + 1.389 + 49.110$
$= 49.805$ m

Check. Level of T_2
$= (-2.25 \times 10^{-4})(100^2) + (2.5 \times 100/100) + 49.110$
$= -2.250 + 2.500 + 49.110 = 49.360$ m (as above)

7.3 Design criteria

7.3.1 Limitation of the rate of change of gradient ($2a$)

This is an empirical method in which $2a = (q-p)/100L$. The value is limited to an arbitrary rate which in practice lies between 5.0×10^{-4} to 5.0×10^{-5}.

Example 7.2 If $2a = -5.0 \times 10^{-4}$, $p = 3\%$ and $q = -2\%$, then
$L = (q-p)/(200a)$
$= -5/(200 \times -2.5 \times 10^{-4}) = 100.00$ m

7.3.2 Minimum equivalent radius (R)

For a line $y = f(x)$, the curvature is
$1/R = [(d^2y/dx^2)/(1 + (dy/dx)^2]^{3/2}$
$1/R \simeq d^2y/dx^2 = 2a$ \hfill (7.8)

Based upon the arbitrary rates above the values of R would vary from 2000 m to 20 000 m.

The main advantage in this method is as a plotting aid at the drawing board stage (see Section 7.4).

7.3.3 Limitation of vertical acceleration

The design speed is usually taken as that speed not exceeded by $x\%$ (say 85%) of the vehicles using the road. The centrifugal acceleration is

$f = v^2/R$ (where f varies from 0.15 to 0.60 m s^{-2})
Then $1/R = 2a = f/v^2$
$$= 3.6^2 f/V^2 = \text{say } 13f/V^2 \qquad (7.9)$$

where v is velocity in m s^{-1} and V is velocity in km h^{-1}. If $V = 60$ km h^{-1} and $f = 0.20$ m s^{-2} then

$$2a = 13 \times 0.20/60^2 = 7.22 \times 10^{-4}$$

If $p = 3\%$ and $q = -4\%$, then

$$L = (q-p)/(100 \times 2a)$$
$$= (-7/100)/-7.22 \times 10^{-4} = 96.95 \text{ m say } 100 \text{ m}$$

7.3.4 Vision (sight) distance (s)

The HMSO publication *Roads in Urban Areas* (RUA 1966) quotes 'To ensure reasonable standards of comfort and appearance and to secure appropriate visibility at *summits*, vertical curves should not be shorter than

$$L = KA \qquad (7.10)$$

where L = curve length, $A = (q-p)$, and K = values given in Design Table II (Table 7.1).

Example 7.3 For a sag curve with a design speed of 60 km h^{-1}, K is given as 20. Then if $p = -3\%$ and $q = 2\%$, then $L = 20 \times (2+3) = 100$ m.

7.4 The scale factor for use of railway curves in the plotting of vertical curves to different scales

The scales are as follows: $1/H$ for horizontal lengths and $1/V$ for vertical heights.

Assuming that the vertical parabolic curve approximates to a circular curve of radius R (R being very large), then the offset height is

$$h \simeq (L/2)^2/2R = L^2/8R \quad \text{(see Fig. 7.6)}$$

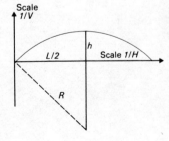

Fig. 7.6

7.4 PLOTTING OF VERTICAL CURVES TO DIFFERENT SCALES

Table 7.1 Design Table II (M) (superseding Design Table II in Layout of Roads in Rural Areas): eye and object height 1.05 m

Carriageway width (m)	Design speed (km h^{-1})	Design capacity (pcus/day Rural Standards)	Minimum stopping sight distance (m)	Minimum overtaking sight distance, single carriageway only (m)	Minimum desirable radius (m)	Absolute minimum radius with maximum 7% superelevation (m)	Minimum radius for curves without transitions (m)	Stopping sight distance K values for crests[4,5]	Stopping sight distance K values for sags[3,5]	Overtaking sight distance K values for crests on single carriageways[5]
(1)	(2)	(3)	(4)	(5)	(6)*	(7)	(8)	(9)	(10)	(11)
Dual 14.60 m	120	66 000	300	—	960*	510	1500	105	75	—
Dual 11.00 m	120	50 000	300	—	960*	510	1500	105	75	—
Dual 7.30 m	120	33 000	300	—	960*	510	1500	105	75	—
10.00 m	100	15 000	210	450	660*	350	1350	50	50	240
7.30 m	100	9 000	210	450	660*	350	1350	50	50	240
7.30 m	80	9 000	140	360	420*	230	1200	25	30	150
—	60	—	90	270	240*	130	600	10	20	90

Notes
1) This table was issued with Technical Memorandum T8/68 in November 1968.
2) * This value amends that given in T8/68.
3) The stopping sight distance for sags is that required for headlamp beams to show up objects on the carriageway.
4) Where economically practicable the K values for crests should be increased preferably to 175 m on rural motorways.
5) To determine length of vertical curve in metres multiply algebraic difference in gradients by K value given in columns (9), (10) or (11) for the appropriate design speed.

and
$$R = L^2/8h \qquad (7.11)$$

If the scale for the plotting of the horizontal lengths is $1/H$ and the scale of the plotting of the vertical offsets (h) is $1/V$ then the actual radius is

$$R' = kR = (HL)^2/8Vh$$
$$= H^2/V \times L^2/8h$$

then $\quad k = H^2/V \qquad (7.12)$

The number on the railway curve represents the radius R in mm (or inches). Thus to use the curve for horizontal alignment, $R'' = HR$ and with the compound scales,

$$R' = \text{(radius in metres)}$$
$$= R'' \times (H/V)$$
$$= HR \times (H/V)$$
$$= R \times H^2/V \quad \text{i.e. } nH^2/1000\,V \qquad (7.13)$$

where n is the number on the curve.

Example 7.4 If $p = 2\%$, $q = -2\%$ and $L = 50$ m, $H = 500$ and $V = 50$, then

$$R' = 1/2a = 100L/(q-p)$$
$$= 100 \times 50/4 = 1250.00 \text{ m (actual value)}$$

If $\quad L = 50.000$ m then at a scale of $1/500$,

$$L' = 50.000/500 = 0.100 \text{ m}$$
$$h = bL/2 - aL^2/4 \quad \text{(at the midpoint)}$$
$$= 0.500 - 0.250 = 0.250 \text{ m}$$

At a scale of $1/50$, h becomes

$$h' = 0.250/50 = 0.005 \text{ m}$$

Then $\quad R = L^2/8h$
$$= 0.100^2/(8 \times 0.005) = 0.250 \text{ m}$$
$$n = 250$$

Check. $\quad R' = R \times H^2/V$
$$= 0.250 \times 500^2/50 = 1250.00 \text{ m}$$

7.5 Sight distances (s)

Formulae relating the length of the vertical curve (L) to the sight distance (s) are given as follows.

7.5.1 Summits

1) $s > L$ $L = 2s - 400(h_1 + h_2)/(p-q)$ (7.14)
2) $s = L$ $L = 400(h_1 + h_2)/(p-q)$ (7.15)
3) $s < L$ $L = s^2(p-q)/[200(\sqrt{h_1} + \sqrt{h_2})^2]$ (7.16)

where h_1 is the height of the eye above the road on the first straight and h_2 is the height of the object above the road on the second straight.

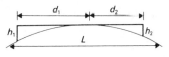

Fig. 7.7

Example 7.5 Given $h_1 = 1.500$ m, $h_2 = 1.000$ m, $p = 2\%$, $q = -3\%$, $s = 200$ m $(s < L)$, find L.

With reference to Fig. 7.7,

$$h_1 = ad_1^2 \text{ then } d_1 = (h_1/a)^{1/2}$$
$$h_2 = ad_2^2 \text{ then } d_2 = (h_2/a)^{1/2}$$
$$s = d_1 + d_2$$
$$= (h_1/a)^{1/2} + (h_2/a)^{1/2} = (1/\sqrt{a})(\sqrt{h_1} + \sqrt{h_2})$$
$$s^2 = (1/a)(\sqrt{h_1} + \sqrt{h_2})^2$$
$$= 200 L (\sqrt{h_1} + \sqrt{h_2})^2/(p-q)$$
$$L = s^2(p-q)/[200(\sqrt{h_1} + \sqrt{h_2})^2] \text{ (equation 7.16 above)}$$
$$= 200^2 \times 5/(200 \times 4.949) = 202.06 \text{ m}$$

7.5.2 Valley curves

Underpasses with clearance (H):

For $s > L$ $L = 2s - 800[H - (h_1 + h_2)/2]/(q-p)$ (7.17)
     $s \leqslant L$ $L = s^2(q-p)/[800(H - (h_1 + h_2)/2)]$ (7.18)

Sight distances related to the length of beam of a vehicle's headlamp:

For $s > L$ $L = 200(s\theta + h)/(q-p)$ (7.19)

where θ is the inclination of the beam (in practice $\theta = 1°$) and h is the height of beam above road ($h = 0.750$ m). Using these values,

$$L = (3.5s + 150)/(q-p) \quad (7.20)$$
For $s \leqslant L$ $L = s^2(q-p)/(3.5s + 150)$ (7.21)

Example 7.6 The sag vertical curve between gradients of 3 in 100 downhill and 2 in 100 uphill is to be designed on the basis that the headlamp sight distance of the car travelling along the curve equals the minimum permitted safe stopping distance at the maximum permitted car speed. The headlamps are 750 mm above the road surface and their beams tilt upwards at an angle of 1° above the longitudinal axis of the car. The minimum safe stopping distance is 150 m. Calculate the length of the curve, given that the length of the curve is greater than the sight distance. (LU)

The values give $p = -3\%$, $q = 2\%$ and $s = 150.00$ m. Then by equation 7.21,

$$L = s^2(q-p)/(3.5s+150)$$
$$= 150.000^2(2+3)/(3.5 \times 150.000 + 150)$$
$$= 166.67 \text{ m}$$

7.6 Setting out data

Gradients A longitudinal section is generally required in order that the mean gradients (p and q) may be derived. The mean values must take into account many factors but the main concern must be the quantities of earth required for cut and fill. To obtain the gradients given the minimum amount of data, levels are taken at four points A, B, C and D as shown in Fig. 7.8.

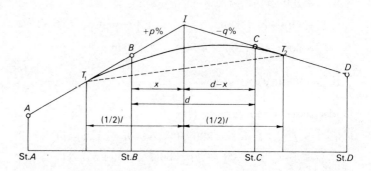

Levels taken at stations A, B, C and D

Fig. 7.8 Setting out data

Chainage and levels of I, T_1 and T_2

Level of I (y_i) = level of B $(y_b) + px/100$
$\qquad$ = level of C $(y_c) + q(d-x)/100$

Solving these equations simultaneously gives the value of x.

Chainage of I (C_i) = chainage of B $(C_b) + x$

Chainage of T_1 (C_{T1}) = chainage of I $(C_i) - L/2$ $\qquad$ (7.22)

Chainage of T_2 (C_{T2}) = chainage of I $(C_i) + L/2$ $\qquad$ (7.23)

Level of T_1 (y_{T1}) = level of I $(y_i) - pL/200$ $\qquad$ (7.24)

Level of T_2 (y_{T2}) = level of I $(y_i) - qL/200$ $\qquad$ (7.25)

7.6.1 Computation of levels on the curve

With reference to Fig. 7.9, conform to the basic equation $y = ax^2 + bx + c$ to obtain the level on the curve at a point P by applying the distance x_p from the tangent point T_1 (i.e. $x_p = (C_p - C_{T1})$) into the equation. Note that $b \equiv p\%$ and c = the level of the tangent point $T1$ (y_{T1}). Then

$$y_p = ax_p^2 + p\% \text{ of } x_p + y_{T1}$$

The value of (a) is obtained from the length (L) or vice versa.

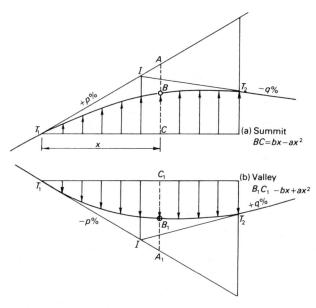

Fig. 7.9 Levels in the curve

Working from first principles the following process is recommended.

1) Compute the levels on the tangent successively for each point in turn; as the standard chord is constant, then a constant difference in level is applied. Thus
   Level on the tangent at a
   = Level of $T_1 + p\%$ of x_a
   Level on the tangent at b
   = Level on the tangent at $a + p\%$ of c_s
2) Compute the component ax^2 and add/subtract to the level of the tangent
3) Compute the values of Δ' and Δ'', as given below (see p. 264) to check the arithmetic.

The whole process is then tabulated as follows.

Example 7.7 Given $p = 3\%$, $q = -4\%$, $L = 300$ m, level of $I = 23.634$ m at chainage 1264.37 m, standard chord 50 m. Then

$$a = (q-p)/(200\ L) = -7/60\,000 = -1.167 \times 10^{-4}$$
$$y_{T1} = 23.634 - (0.03 \times 150) = 19.134$$
$$y_{T2} = 23.634 - (0.04 \times 150) = 17.634$$
$$C_{T1} = 1264.37 - 150.00 = 1114.37$$
$$C_{T2} = 1264.37 + 150.00 = 1414.37$$

Tabulating the levels,

Point	Chainage	x	Level on tangent	ax^2	Level on curve	Δ'	Δ''	
T_1	1114.37		19.134		19.134			
			1.069			0.921		
a	1150.00	35.63	20.203	−0.148	20.055		−0.128	
			1.500			0.793		
b	1200.00	85.63	21.703	−0.855	20.848		−0.584	
			1.500			0.209		
c	1250.00	135.63	23.203	−2.146	21.057		−0.583	constant
			1.500			−0.374		
d	1300.00	185.63	24.703	−4.020	20.683		−0.584	
			1.500			−0.958		
e	1350.00	235.63	26.203	−6.478	19.725		−0.582	
			1.500			−1.540		
f	1400.00	285.63	27.703	−9.518	18.185		−0.989	
			0.431			−0.551		
T_2	1414.37	300.00	28.134	−10.500	17.634			

Note that the final check on the curve level at T_2 proves that the tangent levels have been worked out correctly whilst the *second difference check* proves that the levels on the curve have been computed correctly.

In all cases the value of a will have been derived from L, p and q. The quantity L will probably have been derived relative to the sight distance (s) or from the drawing board design.

7.6.2 The second difference (Δ'') check on the vertical curve levels

Given the basic equation $y = ax^2 + bx + c$, let the length of the standard chord be d. Then

$$x_2 = x_1 + d \quad \text{and} \quad x_3 = x_1 + 2d$$

Then
$$y_1 = ax_1^2 + bx_1 + c$$
$$y_2 = a(x_1 + d)^2 + b(x_1 + d) + c$$
$$= a(x_1^2 + 2x_1 d + d^2) + b(x_1 + d) + c$$
$$\Delta_1' = y_2 - y_1$$
$$= a(2x_1 d + d^2) + bd$$

Similarly,
$$y_3 = a(x_1 + 2d)^2 + b(x_1 + 2d) + c$$
$$= a(x_1^2 + 4x_1 d + 4d^2) + b(x_1 + 2d) + c$$
$$\Delta'_2 = y_3 - y_2$$
$$= a(2x_1 d + 3d^2) + bd$$

The second difference is then
$$\Delta'' = \Delta'_2 - \Delta'_1 = 2ad^2 \quad (7.26)$$

Example 7.8 If $a = -4.34 \times 10^{-5}$, $d = 20$ m, then $2ad^2 = -0.035$.

Chainage	Level on curve	Δ'	Δ''
0	40.640		
		0.183	
20	40.823		-0.035
		0.148	
40	40.971		-0.035
		0.113	
60	41.084		-0.035
		0.078	
80	41.162		-0.034
		0.044	
100	41.206		-0.035
		0.009	
120	41.215		-0.035
		-0.026	
140	41.189		-0.034
		-0.060	
160	41.129		

Note that as the curve is normally set out using through chainage, then the chainage difference will not be constant for the first and last sub-chords, i.e.
$$d_1 \neq d_2 = d_3 = d_4 = d_{n-1} \neq d_n$$

Example 7.9 As part of a dual highway reconstruction scheme, the following line of levels was taken at given points on the existing surface.

	Reduced level	Chainage
A	31.891	6 +32.46
B	33.263	6 +78.18
C	32.702	7 +77.24
D	31.635	8 +30.58

If the curve, based on a simple parabola, is designed to give a rate of change of gradient of 2.5×10^{-4}, calculate: (a) the length of the curve L; (b) the chainage and level of the intersection point (I); (c) the chainages and levels of the tangent points; (d) the level of the points on the curve (based on standard chords of 20 m); (e) the length of the line of sight (s) to a similar vehicle of a driver 1.05 m above the road surface ($s < L$).

The solutions refer to Fig. 7.10.

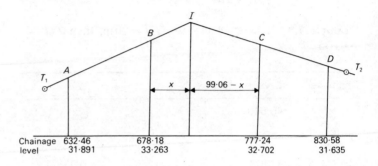

Fig. 7.10

(a) Gradient $AB = (33.263 - 31.891)$ in $(678.18 - 632.46)$
$$= 1.372 \text{ in } 45.72, \text{ i.e. } 1 \text{ in } 33.3 \, (+3\,\%)$$
$CD = (31.635 - 32.702)$ in $(830.58 - 777.24)$
$$= -1.067 \text{ in } 53.34, \text{ i.e. } 1 \text{ in } 49.99 \, (-2\,\%)$$

By equation 7.5, length of curve is
$$L = (3+2)/(100 \times 2.5 \times 10^{-4}) = 200.00 \text{ m}$$

(b) Level of I
$$= 33.263 + 3x/100 = 32.702 + 2(99.06 - x)/100$$

Solving for x,
$$x = 28.40 \text{ m}$$
$$99.06 - x = 70.66 \text{ m}$$

Level of $I = 33.263 + 0.03 \times 28.40 = 34.115$ m
$$= 32.702 + 0.02 \times 70.66 = 34.115 \text{ m (check)}$$

Chainage of I = chainage of $B + x$
$$= 678.18 + 28.40 = 706.58 \text{ m}$$

(c) Chainage of $T_1 = 706.58 - 100.00 = 606.58$ m
Chainage of $T_2 = 706.58 + 100.00 = 806.58$ m
Level of $T_1 = 34.115 - (3 \times 200/200) = 31.115$ m
Level of $T_2 = 34.115 - (2 \times 1) = 32.115$ m

(d) Setting out data

Point	Chainage	x	Level on tangent	ax^2	Level on curve	Δ'	Δ''
T_1	606.58		31.115		31.115		
			0.403			0.380	
a	620.00	13.42	31.518	−0.023	31.495		0.103
			0.600			0.483	
b	640.00	33.42	32.118	0.140	31.978		−0.100
			0.600			0.383	
c	660.00	53.42	32.718	−0.357	32.361		−0.100
			0.600			0.283	
d	680.00	73.42	33.318	−0.674	32.644		−0.100
			0.600			0.183	
e	700.00	93.42	33.918	−1.091	32.827		−0.100
			0.600			0.083	
f	720.00	113.42	34.518	−1.608	32.910		−0.100
			0.600			−0.017	
g	740.00	133.42	35.118	−2.225	32.893		−0.100
			0.600			−0.117	
h	760.00	153.42	35.718	−2.942	32.776		−0.100
			0.600			−0.217	
i	780.00	173.42	36.318	−3.759	32.559		−0.100
			0.600			−0.317	
j	800.00	193.42	36.918	−4.676	32.242		0.190
			0.197			−0.127	
T_2	806.58	200.00	37.115	−5.000	32.115		

Note that
$$\Delta'' = 2ad^2 = 2 \times 1.25 \times 10^{-4} \times 20^2 = -0.100$$

(e) Line of sight (s),

$$h_1 = h_2 = ad^2$$
$$d = (h_1/a)^{1/2}$$
$$s = 2d = 2(h_1/a)^{1/2}$$
$$= 2[1.05/(1.25 \times 10^{-4})]^{1/2}$$
$$= 183.30 \text{ m}$$

Example 7.10 A 6% downgrade on a proposed road is followed by a 1% upgrade. The chainage and reduced level of the intersection point of the grades is 2010.00 m, and 58.62 m AOD respectively. A vertical parabolic curve, not less than 250 m long, is to be designed to connect the two grades. Its actual length is to be determined by the fact that at chainage 2180.00 m, the reduced level on the curve is to be 61.61 m to provide adequate headroom under the bridge at that point. Calculate the required length of the curve and also the chainage and reduced level of its lowest point. (RICS/M)

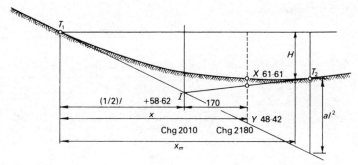

Fig. 7.11

Let $x - L/2 = 170.00$ m (Fig. 7.11)
Tangent level at Y is

$$58.62 - (170.00 \times 0.06) = 48.42 \text{ m}$$

Curve level at $X = 61.61$ m

But $XY = ax^2 = 13.19$ m
$$a = 13.19/x^2$$
$$= (q - p)/200L = 7/200L$$

Thus $13.19/x^2 = 7/200L$
$$L = 7x^2/(200 \times 13.19) = 7x^2/2638$$

Then $x - 7x^2/5276 = 170.00$
$$7x^2 - 5276x - 896\,920 = 0$$

Solving for x,

$$x = (+5276 \pm [5276^2 - (4 \times 7 \times 896\,920)]^{1/2})/14$$
$$= (5276 \pm 1649.9)/14$$
$$= 494.7 \text{ m} \quad \text{or} \quad 259.0 \text{ m}$$
$L/2 = x - 170.00$
$$= 324.7 \text{ m} \quad \text{or} \quad 89.0 \text{ m}$$

Then $L = 650$ m as L must not be less than 250 m.

To find minimum height on the curve,

$$x_m = pL/(p - q)$$
$$= 6 \times 650/7 = 557.14 \text{ m}$$

Chainage of minimum height

$$= 2010.00 + 557.14 - 650/2 = 2242.14 \text{ m}$$

Level of minimum height:

Level of $T_1 = 58.62 + (0.06 \times 325)$
$$= 78.12 \text{ m}$$

Level on tangent at x

$$= 78.12 - (0.06 \times 557.14) = 44.70 \text{ m}$$
$$ax^2 = (7 \times 557.14^2)/(200 \times 650) = 16.71 \text{ m}$$

Level on curve at $x_m = 44.70 + 16.71 = 61.41$ m

Exercises 7.1

1 An uphill gradient of 1 in 100 meets a downhill gradient of 0.45 in 100 at a point where the chainage is 122.880 m and the reduced level is 126.000 m. If the rate of change of gradient is to be 1.18×10^{-4}, prepare a table for setting out the curve at intervals of 20 m.
(ICE Ans. 125.386, 125.551, 125.683, 125.769, 125.807, 125.798, 125.742, 125.724)

2 A rising gradient of 1 in 100 meets a falling gradient of 1 in 150 at a level of 41.600 m AOD. Allowing for headroom and working thickness, the vertical parabolic curve joining the two straights is to be at a level of 41.200 m at its midpoint. Determine the length of the curve and the levels at 20 m intervals from the first tangent point.
(LU Ans. 192.000 m; 40.823, 40.971, 41.084, 41.162, 41.206, 41.215, 41.189, 41.129, 41.034, 40.960)

3 On a straight portion of a new road, an uphill gradient of 1 in 100 was connected to a downhill gradient of 1 in 150 by a vertical parabolic summit curve of length 100 m. A point P, at chainage 11 820.040 m on the first gradient, was found to have a reduced level of 9.024 m, and at a point Q, at chainage 12 000.040 m on the second gradient, the level was 8.990 m.
(a) Find the chainages and reduced levels of the tangent points to the curve.
(b) Tabulate the reduced levels on the curve at intervals of 20 m and at its highest point.
Find the minimum sighting distance to the road surface for each of the following cases.
(c) The driver of a car whose eye is 1.220 m above the surface of the road.
(d) The driver of a lorry for whom the similar distance is 1.830 m. (Take the sighting distance as the length of the tangent from the driver's eye to the road surface.)
(LU Ans. 11 840.00 m; 9.224, 9.391, 9.491, 9.524, 9.491, 9.391, 196.40, 269.60)

4 A falling gradient of 4% meets a rising gradient of 5% at chainage 490.00 m and level 43.284 m AOD. At chainage 470.00 m, the underside of a bridge has a level of 49.108 m. The two gradients are to be joined by a vertical parabolic curve

giving 4.800 m clearance under the bridge. List the levels at 20 m intervals.
(Ans. $L = 80$, 44.884, 44.309, 44.184, 44.509, 45.284.)

5 A rising gradient of 1 in 50 is to be joined by a rising gradient of 1 in 20. The two gradients are to be joined by a vertical curve of parabolic form which is to commence at a reduced level of 28.000 m on the approach gradient, and must pass through a point 40.000 m horizontally from the tangent point, having a reduced level of 29.000 m. Obtain the equation of the curve and calculate (a) the horizontal distance from the tangent point to the intersection point of the two gradients, and (b) levels at intervals of 20 m on the curve showing the 'second difference check'.
(TP Ans. $y = 1.25 \times 10^{-4} x^2 + 0.02x + 28.000$; (a) 60.000 m;
(b) 28.000, 28.450, 29.000, 29.650, 30.400, 31.250, 32.200)

6 T_1, T_3 and T_2 are tangent points on a vertical curve. The gradients of the tangents at these points are $+1$ in 60, $+1$ in 80, and -1 in 100 respectively. If the terminal gradients at T_1 and T_2 meet at the intersection point I of level 30.500 m and the length of the curve is 75.000 m, calculate the length of the curve and the level at T_3. (TP 11.719 m, 30.046 m)

7 In a proposed layout at a junction on a locomotive haulage approaching a pit bottom, a rising gradient of 1 in 100 meets a level gradient at I of level -40.000 m. A vertical curve of length 50.00 m joins these two gradients and then the same curve continues to meet a downhill gradient of 1 in 70. Calculate: (a) the equation of the vertical curve; (b) the total length of the vertical curve; (c) the levels of the tangent points; (d) the amount of 'cutting' required at the two intersection points.
(TP Ans. $-0.0001x^2 + 0.01b - 40.250$, 121.430 m, -40.250, -40.000, -40.511, 0.063, 0.128)

8 Part of a mine-car circuit at a pit bottom, where a falling gradient of 1 in 10 meets a rising gradient of 1 in 50 at an existing level of 145.63 m above an assumed datum, is to be modified by providing a vertical curve. To avoid excessive ripping, the new level at the intersection point is to be 146.83 m. Calculate the length of curve required, the levels at the tangent points, levels on the curve at 10 m intervals, and the filling necessary at each point to bring the circuit to its new level.
(TP Ans. 80.000 m, 149.630, 146.430, Fill 0.075, 0.300, 0.675, 1.200, 0.675, 0.300, 0.075)

Bibliography

SHEPHERD, F. A., *Surveying Problems and Solutions.* Edward Arnold (1968)

Appendix

1 Derivatives of simple functions

$y = ax^n$ $y' = nx^{n-1}$
$y = a/x^n$ $y' = -nax^{-n-1}$
$y = \sqrt{x}$ $y' = 1/(2\sqrt{x})$
$y = 1/x$ $y' = -1/x^2$
$y = \sin x$ $y' = \cos x$
$y = \cos x$ $y' = -\sin x$
$y = \tan x$ $y' = \sec^2 x = 1 + \tan^2 x$
$y = \cot x$ $y' = -\operatorname{cosec}^2 x = -(1 + \cot^2 x)$
$y = \arcsin x$ $y' = 1/\sqrt{(1-x^2)}$
$y = \arccos x$ $y' = -1/\sqrt{(1-x^2)}$
$y = \arctan x$ $y' = 1/(1+x^2)$
$y = \operatorname{arc cot} x$ $y' = -1/(1+x^2)$

The following points should be noted.

1) If $y = uv$ where u and v are functions of x, then

$$y' = u\,dv + v\,du$$

2) If $y = u/v$ then

$$y' = (v\,du - u\,dv)/v^2$$

2 Comparison of plane and spherical trigonometrical formulae

Plane formulae

$$\cos C = \frac{a^2 + b^2 - c^2}{2ab}$$

$$\tan \frac{C}{2} = \left(\frac{(s-a)(s-b)}{s(s-c)}\right)^{1/2}$$

$$a/\sin A = b/\sin B = c/\sin C = 2R$$

Spherical formulae

$$\cos C = \frac{\cos c - \cos a \cos b}{\sin a \sin b}$$

$$\tan \frac{C}{2} = \left(\frac{\sin(s-a)\sin(s-b)}{\sin s \sin(s-c)}\right)^{1/2}$$

$$\sin a/\sin A = \sin b/\sin B = \sin c/\sin C$$

$$\tan(A-B)/2 = \frac{(a-b)}{(a+b)} \cot C/2$$

$$\tan(A-B)/2 = \frac{\sin(a-b)/2}{\sin(a+b)/2} \cot C/2$$

$$\tan(A+B)/2 = \cot C/2$$

$$\tan(A+B)/2 = \frac{\cos(a-b)/2}{\cos(a+b)/2} \cot C/2$$

Right-angled triangle: $C = 90°$

$$\sin a = \sin A . \sin c = \cot B . \tan b$$
$$\sin b = \sin B . \sin c = \cot A . \tan a$$
$$\cos c = \cos a . \cos b = \cot A . \cot B$$
$$\cos A = \cos a . \sin B = \tan b . \cot c$$
$$\cos B = \cos b . \sin A = \tan a . \cot c$$

3 Series

1 Binomial expansion

$$(a+x)^n = x^n + nx^{n-1}a + \frac{n(n-1)}{2!}x^{n-2}a^2 + \frac{n(n-1)(n-2)}{3!}x^{n-3}a^3 + \ldots$$
$$+ \frac{n(n-1)(n-2)\ldots(n-r+1)}{r!}x^{n-r}a^r + \ldots$$

2 Taylor's series for a single variable

$$f(x+h) = f(x) + hf'(x) + \frac{h^2}{2!}f''(x) + \frac{h^3}{3!}f'''(x) + \ldots + \frac{h^n}{n!}f^n(x) + \frac{h^{n+1}}{(n+1)!}f^{n+1}(x) \ldots$$

3 Exponential series

$$e^x = 1 + x + \frac{x^2}{2!} + \frac{x^3}{3!} + \frac{x^4}{4!} + \ldots \quad e = 2.718\,281\,8285$$

4 Trigonometric series (θ in radians)

$$\sin \theta = \theta - \theta^3/3! + \theta^5/5! - \theta^7/7! + \ldots$$
$$\cos \theta = 1 - \theta^2/2! + \theta^4/4! - \theta^6/6! + \ldots$$
$$\tan \theta = \theta + \theta^3/3 + 2\theta^5/15 + 17\theta^7/315 + 62\theta^9/2835 + \ldots$$

$$\sin^{-1}\theta = \theta + \frac{1}{2}\cdot\frac{\theta^3}{3} + \frac{1\times 3}{2\times 4}\cdot\frac{\theta^5}{5} + \frac{1\times 3\times 5}{2\times 4\times 6}\cdot\frac{\theta^7}{7} + \ldots$$

$$\cos^{-1}\theta = \pi/2 - \sin^{-1}\theta$$
$$\tan^{-1}\theta = \theta - \theta^3/3 + \theta^5/5 - \theta^7/7 + \ldots$$

Index

Accidental errors, 2
accuracy, 1, 112
adjustments, 13, 19
adjustments of observations, 19
   braced quadrilateral, 87
   by variation of coordinates, 43, 54, 73
   conditional equations, 32
   curvature and refraction, 142
   EDM networks, 37
   'least squares', 13, 19, 21, 32
   levelling networks, 26
   observational equations, 19
   spherical excess, 150
   station observations, 27
AGA Geotronics, 122, 148
Airy's spheroid, 164, 168, 173, 175
alignment,
   horizontal, 208
   vertical, 254
Allan, A. L., 45, 183
amplitude, 124
amplitude method (gyro), 100, 105
angle conditions, 87
angle of swing, 91
angular distortion $(t-T)$, 177
application of superelevation, 263
arbitrary meridian, 82
arc/chord approximation, 188
arc length, 188
arc to chord 'correction', 178
area of a spherical triangle, 149
arithmetic mean, 3
atmospheric
   pressure, 121
   refraction, 121
auxiliary base measurement, 116, 117
average value, 3

Back angle, 201, 233, 236
Barrell, H., 121, 148
base line (auxiliary), 116
bearings, 43, 48, 52, 63
Bernoulli lemniscate, 229
Bird, R. G., 44
Briggs, N., 13, 45
British Standard 2846 (1957), 46
Burnside, C. D., 148

Calculus of errors, 15, 130
calibration (EDM), 127
cant, 224
carrier waves, 120
Cassini projection, 169
cartesian coordinates, 184, 205, 226
Cauchy's equation, 120
central meridian, 163
chainage, through, 187
chord and offset, 201
circles, great and small, 149
Clark, D., 140, 145, 148, 165, 183
Clark's equation, 140, 145, 165
closing errors, 1
clothoid, 226
coefficient of refraction, 136
comparison of direct and indirect methods, 41
compound curves, 210
concrete slab, 38
conditional equations, 19, 32
confidence intervals, 9
constant, gyro, (E) 106
constant errors, 2
control measurements, 111
convergence of meridians, 162, 180
Cooper, M. A. R., 46, 148, 183
coordinates,
   around the curve, 205
   cartesian, 184, 205, 226
   geographical, 158, 175
   National Grid, 158, 163, 175
   spherical,
   corrections', 141, 142
correlation, 90
correlatives, 32
correlative normal equations, 33, 35, 37
Criswell, 230
criteria of design, 224, 257
criterion of negligibility, 13, 94
crossed quadrilateral, 87
cubic parabola, 229
curvature,
   and refraction, 142
   minimum, 224
   of the earth, 141
   of the path of the EDM wave, 135
curves,

Curves *(contd.)*
  circular, 184
  compound, 210,
  coordinates, 205
  fixed, floating, free, 208
  reverse, 217
  special problems of, 190
  transition, 223
  vertical, 254
cyclic error, 127, 128

Degree of curve (D), 184
design criteria
  for transition curves, 224
  for vertical curves, 257
deviation, standard, 5, 6
differential displacements, 43
direct method of adjustment, 32, 41
'direction' method, 76
distances,
  horizontal (H), 111, 129
  measured (D), 129
  spheroidal (S), 141, 143
doubtful observations, 7

Eccentricity (e), 164
EDM
  accuracy, 127
  calibration, 127
  network, 37
electromagnetic distance measurement (EDM), 111
ellipse, 22
equation
  of a circle, 184
  of a straight line, 184
equipotential surface, 163
equivalent circle, 211
errors,
  closing, 1
  constant, 2
  periodic, 2
  random (accidental), 2
  residual, 2
  standard, 5, 6
  systematic, 2
  true, 1
Essen, L., 148
eye-object 'correction', 140

Flattening (f), 164
Froome, K. D., 148

Gaussian notation, 3
geodesy, 141
geodesics, 178
geodimeter, 124
geographical coordinates, 158, 162, 175
geoid, 163
geometry of the curve, 188

gradients, 262
great circle, 149
grid coordinates, 163, 174, 175, 181
grid north, 47
gyroscopic
  observations, 47
  theodolite,
    floating type, 97
    suspended type, 97, 99

Hayford's spheroid, 164
heighting, trigonometrical, 139
Helliwell, E. G., 134, 148
highway alignment, 208
Highway Transition Curve Tables, 225, 230, 233
HMSO, 183
Hodges, D. J., 109, 136, 148
Hollwey, J. R., 45, 183
horizontal
  length (H), 111, 123, 130
  scale, 184
humidity, 120

Index, refraction, 120, 142
indirect method of adjustment, 19
Institution of Municipal Engineers, 236
intersection, 47, 48
  multi-ray, 51
  semi-graphic, 51
  two-ray, 48
  variation of coordinates, 55

Jameson, A. H., 46, 183

K (coefficient of vertical refraction), 136, 142
k (zero error), 127, 128

Latitude (ψ), 158, 161, 162, 165, 168, 169, 175, 180
least squares, 13, 19
Lee, L. P., 183
Legendre's theorem, 150
levelling circuits, 26
levels on vertical curves, 263
L'Huilier's equation, 150
Lilley, J. E., 109
limitation of vertical acceleration, 258
local scale factor, 141, 178
location of a curve, 194
lune, 149

Maling, D. H., 183
matrix, 20
  variance–covariance, 21, 22
maximum/minimum height on the vertical curve, 256
Maynes, J. H. B., 183
mean value, 3
measures of precision, 4

Mekometer, 120
Mercator, transverse projection, 164, 169, 171, 183
meridians, 47
   convergence of, 162
minimum curvature, 224
minimum equivalent radius, 257
mnemonics, Napier's, 154
modified gyroscopic theodolite, 98
most probable value, 3, 19
Murchinson, D. E., 46

Napier's mnemonics, 154
National Grid, 163, 178
normal equations, 19, 20
north: assumed, grid, magnetic, true, 47

Observational equations, 19
optical distance measurement, 111
Ordnance Survey, 169
orientation, 47
oscillation period of the gyro, 99

Peter's approximation, 7
phase comparison, 120
polar
   coordinates, 205
   triangle, 154
pole, 149
precession, 98
precision, 1
Price, W. E., 148
probability density equation, 13
projection tables, 169, 170, 171
projections,
   Cassini, 167
   transverse Mercator, 169
propagation of errors,
   random, 12
   systematic, 11
proportional error, 128

Radial acceleration, 225
radius of curvature, 223, 233
railway curves, 184
Rainsford, H. F., 46
rate of change of gradient ($2a$), 254
redundant observations, 19
reference spheroids, 164
refractive index, 120
rejection criteria, 7
resection, 47, 65
   semi-graphic, 71
   three-point, 67
   variation of coordinates, 73
reversal point method, 100
Richardus, P., 46
right-angled spherical triangle, 154
Robbins, A. A., 135, 148
Royal School of Mines, 100
RUA (roads in urban areas), 259

Saastamoinen, J., 136, 148
sample
   precision, 5
   size, 10
satellite stations, 82
scale factor, 91
   railway curves, 258
Schofield, M. W., 46
Schuler's mean, 100, 105
Schwendener, H. R., 103, 148
second difference check, 264, 265, 267
semi-graphic
   intersection, 51
   linear, 58
   resection, 71
setting out data,
   circular, 198
   transition, 234
   vertical, 262
shift (s), 227
side equations, 87
sight distances, 258, 259, 260
simple parabola used in vertical curves, 254
small circles, 149
Smith, R. C. H., 110
spherical
   trigonometrical formulae, 271
   trigonometry, 149
   excess, 150
spheroid, 163
spheroidal coordinates, 164
standard deviation, 5, 6
standard error, 5, 6
   an observation of weight (w_i), 9
   of a single observation, 6
   the mean, 6
   the weighted mean, 9
station adjustment, 28
Student-t, 9
subtense bar, 111
superelevation, 223
   application, 236
   on roadways, 224

Tacheometric constants, 30
tangent length, 188, 228
tangential angles, 227
tape zero, 99
($t - T$) 'correction', 177, 178
Thomas, T. L., 100, 109
Thornton-Smith, G. J., 110
three-point resection, 65, 67
through chainage, 187
transformation of coordinates, 90
transit method, 102
transition curves, 223
traverse network, 37
trigonometrical heighting, 111

trigonometry, spherical, 149
two-point resection, 65

Unbiased estimate, 22
Uren, J., 148

Variance, 5
variance–covariance matrix, 21, 24
variation of coordinates, 43
　intersection, 55
　linear, 60
　resection, 73

vertical curves, 254
　design criteria, 257
　properties, 255
　second difference check, 264
　setting out data, 262
　slight distances, 260

Weight matrix, 21
weighted observations, 8
Weisbach triangle, 92

Zero error (EDM), 127, 128, 129
zero position of gyro-tape, 99